Viktoria Wegewitz

If females can reproduce by themselves, why are there males?

Viktoria Wegewitz

If females can reproduce by themselves, why are there males?

The role of males in an androdioecious system

Südwestdeutscher Verlag für Hochschulschriften

Impressum / Imprint
Bibliografische Information der Deutschen Nationalbibliothek: Die Deutsche Nationalbibliothek verzeichnet diese Publikation in der Deutschen Nationalbibliografie; detaillierte bibliografische Daten sind im Internet über http://dnb.d-nb.de abrufbar.
Alle in diesem Buch genannten Marken und Produktnamen unterliegen warenzeichen-, marken- oder patentrechtlichem Schutz bzw. sind Warenzeichen oder eingetragene Warenzeichen der jeweiligen Inhaber. Die Wiedergabe von Marken, Produktnamen, Gebrauchsnamen, Handelsnamen, Warenbezeichnungen u.s.w. in diesem Werk berechtigt auch ohne besondere Kennzeichnung nicht zu der Annahme, dass solche Namen im Sinne der Warenzeichen- und Markenschutzgesetzgebung als frei zu betrachten wären und daher von jedermann benutzt werden dürften.

Bibliographic information published by the Deutsche Nationalbibliothek: The Deutsche Nationalbibliothek lists this publication in the Deutsche Nationalbibliografie; detailed bibliographic data are available in the Internet at http://dnb.d-nb.de.
Any brand names and product names mentioned in this book are subject to trademark, brand or patent protection and are trademarks or registered trademarks of their respective holders. The use of brand names, product names, common names, trade names, product descriptions etc. even without a particular marking in this works is in no way to be construed to mean that such names may be regarded as unrestricted in respect of trademark and brand protection legislation and could thus be used by anyone.

Coverbild / Cover image: www.ingimage.com

Verlag / Publisher:
Südwestdeutscher Verlag für Hochschulschriften
ist ein Imprint der / is a trademark of
AV Akademikerverlag GmbH & Co. KG
Heinrich-Böcking-Str. 6-8, 66121 Saarbrücken, Deutschland / Germany
Email: info@svh-verlag.de

Herstellung: siehe letzte Seite /
Printed at: see last page
ISBN: 978-3-8381-3696-7

Zugl. / Approved by: Tübingen, Eberhard Karls Universität, Diss., 2009

Copyright © 2013 AV Akademikerverlag GmbH & Co. KG
Alle Rechte vorbehalten. / All rights reserved. Saarbrücken 2013

Acknowledgements

This thesis work was carried out under the guidance of Dr. Adrian Streit in the Department of Evolutionary Biology chaired by Prof. Dr. Ralf Sommer at the Max-Planck-Institut für Entwicklungsbiologie in Tübingen, Germany.

I thank Prof. Dr. Ralf Sommer for giving me this great chance to work as a PhD student in his department.

I would like to thank Prof. Dr. Ralf Sommer and Prof. Dr. Nico Michiels to be examiners of my doctoral work at the Fakultät für Biologie, der EBERHARD KARLS UNIVERSITÄT TÜBINGEN.

There are no words describing how thankful Adrian I am for his endless support and for sharing his extensive knowledge with me.

Also thanks to Prof. Dr. Hinrich Schulenburg for good collaboration with the experiments.

I thank this whole department IV for providing this good working atmosphere.

Thank you Dr. Robbie Rae for proof-reading the manuscript.

I thank Heike Haussmann for cheering me up and for driving me around.

Thanks to Armin Mayer, who wrenched myself away from daily life more than a couple of times.

And, of course, thanks to my mother, Ulrike Wegewitz, who always kept me grounded.

Table of Contents

1. Introduction

    1.1 Evolution of sex……………………………………………….. 4
    1.2 Sexual conflict…………………………………………………….. 8
    1.3 Evolution of androdioecy………………………………………. 9
    1.4 Nematode - basic morphology………..………………………. 11
    1.5 Phylogenetic position of *Caenorhabditis elegans*……………. 12
    1.6 Common aspects – lifestyle of *Caenorhabditis elegans*………. 13
    1.7 Why are there still males?.. 15
    1.8 Outcrossing rates in natural populations……………………. 16
    1.9 Mating behavior………………………………………………. 18

2. Aim of the thesis…………………………………………………… 21

3. Results and Discussion

    3.1 Experimental insight into the proximate causes of male persistence variation among two strains of the androdioecious *Caenorhabditis elegans* (Nematoda) (Synopsis)………………………………………………..…… 22

    3.2 Do males facilitate the spread of novel phenotypes within populations of the androdioecious nematode *Caenorhabditis elegans* (Synopsis)……………………… 25

Bibliography… ………………………………………………………….. 28

Appendix
    Summary/Zusammenfassung……………………………………. 33
    Publications……………………………………………………… 36

1. Introduction

The philosopher Platon describes the mythos of the round man. In the beginning there existed three types of gender, the male one arose from the sun, the female from the earth and the mixture of both from the moon. These round men possessed four hands and feet each and two faces on the head. If they wanted to move faster they leaned on eight extremities to rotate around. They were not only fast but also strong and they wanted to pave the way to the sky to attack the gods. Zeus did not want to kill them but on the other hand wanted to penalize the round men. He decided to cut them into two. Since then both parts walk upright on to two legs, and both parts have the desire to unify with the other part. Zeus moved the genitals to the front such that if a male meets a female they can produce progeny. This impulse of the two halves to join together is called love (erôs) (after Susemihl, 1855.) Love lets one overcome the feeling of isolation and at the same time allows one, to be itself and to keep his integrity. In love it comes to the paradoxon, that two people become one and despite of that remain two (after Fromm, 2001).

1.1 Evolution of sex

Reproductive strategies can be split into sexual and asexual reproduction. Asexual reproduction is the process by which an organism creates a genetically-similar or identical copy of itself, without the contribution of genetic material of an other individual. Examples include fission of bacteria, budding yeast or Hydras, which can split into two or more individuals. However, some species, such as Hydra and Jellyfish, can reproduce sexually or asexually. Many plants are capable of vegetative reproduction—reproduction without seeds or spores—but can also reproduce sexually. Also by conjugation bacteria may exchange genetic information. Other ways of asexual reproduction include parthenogenesis, fragmentation and spore formation that involves only mitosis. Parthenogenesis is the clonal growth and development of embryos or seeds (Wehner & Gehring, 2007).

Sexual reproduction is a process where two organisms build haploid gametes, which then fuse to create a new organism with a combination of the genetic material (Wehner & Gehring, 2007). A special kind of sexual strategies is hermaphroditism where one organism comprises both female and male gender. They can either self-fertilize or cross-fertilize each other. If facultative functional males are produced which successfully fertilize hermaphrodites, it is called androdioecy. The other way around if there is a functional female which can be fertilized by the hermaphrodites, it is called gynodioecy.

One mysterium on earth is why is there sex and how did it arise? How could nature invent females, which produce gametes equipped to nourish an embryo and at the same time evolve a male to produce mobile sperm cells? And how could it happen that the gametes contain half the chromosomal number in comparison to somatic cells (Göldenboog, 2006)?

At first sight sexual reproduction with separate females and males appears paradoxical compared with a population of non-sexual females. Females in a bisexual population have double the costs because the progeny is split into males and females, where only the female can propagate again. In an asexual organism all progeny is female and can reproduce. This implies that with each generation, an asexual population can grow more rapidly (Maynard Smith, 1978). An additional cost is that males must find females, in this process often sexual selection favors traits that reduce the fitness of the individuals. What are the offsetting benefits? Probably it is a combination of several aspects, which now will be discussed.

I) Sex creates variability between siblings (Weismann, 1889)

George C. Williams „Sex and Evolution"(Williams, 1975) talks about the idea that sex introduces genetic variation for organisms to survive in changing environmental conditions. The idea is that organisms, which instead of giving rise to a large number of offspring with one genotype, produce offspring with a range of types, some of which will have a better chance to survive in novel environments. Critics of this theory claim that sex is most common in stable environments like e.g. the tropic whereas if the environments is unstable such as at high altitudes or in small body waters asexual reproduction prevails.

II) Sex creates novel genotypes more rapidly (Fisher, 1930)

When two genomes are combined by sex, advantageous alleles can be combined in a chromosome within a few generations by recombination. In the case of positive epistasis among the involved alleles this can lead to fitter progeny (i.e., higher reproductive rate). Ronald Fisher suggests that sex might facilitate the spread of advantageous alleles by allowing them to escape their genetic background, which might be deleterious. Most mutations are deleterious and, i.e. if a disadvantageous mutation lays on a chromosome with a gene that is important for the organism, the disadvantageous gene will be protected by the advantageous gene and cannot be diminished by natural selection. The advantageous genes are disengaged from their chromosomal context through sex and recombination. In the best case beneficial genes are on one chromosome and deleterious genes on the other which leads to the according effect that sex bring together not only positive but also deleterious mutations. Through negative epistasis where natural selection can work on, these mutations are removed from the genome. In asexual populations deleterious mutations will accumulate and the population eventually will become unable to reproduce (i.e. go extinct). This is known as Muller's Ratchet (Muller, 1964). On the other hand sex should be disadvantageous when positive genes are embedded in „good company". In this case, asexual reproduction should be the best way to reproduce since favourable gene combinations are not broken apart (Kondrashov, 1988).

One common way of testing for epistasis is blotting the log fitness against the number of mutations relative to a reference strain. If there is no epistasis, the fitness effects of the individual mutations contribute multiplicatively to fitness, and log fitness decreases linearly with increasing number of mutations. If epistasis is positive the fitness curve will be less than linear, and a more than linear decrease suggests negative epistasis. (Bonhoeffer et al., 2004)

There are some aspects that contradict the assumption, that sex increases variability and that way allows evolution to proceed faster. First sexual reproduction and recombination do not always increase genetic variation. Second, even when they do, it is not clear why greater variation should generally be adaptive. Third recombination also destroys favourable combinations. (Bonhoeffer et al., 2004)

III) The Red Queen hypothesis states that parasites are constantly evolving, and in this situation sex helps to coevolve and adapt to these new environments. The result is a constant arms race. If the interaction between host and parasite persists over time, then this results in repeated reciprocal adaptations among the two antagonists. Here, sex may help the host to respond rapidly to parasites, which are usually assumed to evolve faster because of their relatively larger population size, shorter generation time and often haploid genome. Sex helps increase the resistance to parasites because individuals need to continually adopt to rapid changes of the environment to stay ahead of their parasites. (Hamilton et al., 1990). This is known as the red queen hypothesis, taken from Lewis Carrolls "Alice in Wonderland" where the red queen tells Alice that she has to run as fast as she can to stay at the same place, meaning she has to run to keep up with the changing environment (Carroll, 1865).

IV) It has been proposed that one of the advantages of sex is that during meiosis recombination provides an opportunity for repair of double-strand damages in chromosomes (Cox, 2001).

1.2 Sexual conflict

The two genders have different interests, which conflict with each other. This causes sexual conflict and results in sexual selection. There are two kinds of sexual selection: male competition and female choice. Males compete for access to females and their eggs. A higher reproductive success and therefore a higher fitness can be reached with more successful inseminations (Stockley et al., 1995). The females´ ability to produce gametes is much more restricted, because eggs are more costly than sperm. Therefore, females are usually choosier than males (Bateman, 1948). There are different optima in mating frequency and partner choice in the sexes. This causes sexual conflict and results in sexual selection and coevolution of many sex-related traits (Parker, 1970). As a consequence, these conflicts of interests lead to an amazing diversity of behaviour and other traits in nature (Johnstone & Keller, 2000).

For males multiple mating has multiple fitness advantages but mating itself does not ensure paternity because the sperm of a competitor can outcompete a particular male's sperm. Because of that evolution favours males that manage to prevent females from mating more often or to make the female use its sperm and not the competitor's sperm. These strategies are often harmful to the female and represent a cost of mating (harmful impact).

There are different hypotheses about the harmful impact of males to females. For example, the pleiotropic harm hypothesis where male increases its fitness by manipulating the female. Costs that result from male harm are side effects and not themselves selectively advantageous. Lower female fecundity costs are outweighed through reproductive benefits such as increased paternity (Morrow et al., 2003). Another example is the adaptive harm hypothesis where harming females happens on purpose. Males manipulate females to adaptively modify their life history towards the males benefit. Lowering the females remating rate is one possibility to respond to harmful substances like seminal toxins (Johnstone

& Keller, 2000). Another possibility is that females reallocate resources from maintenance to current reproduction, because their future reproductive value is decreased after injury (Lessells, 2005). Because the individual expects an early death it tries to produce as much offspring as possible (Michiels, 1998). On the other hand females, which are able to resist or compensate male harm are strongly favoured by selection. Antagonistic coevolution occurs if the sexes reciprocally adapt to each other (Gavrilets et al., 2001). There is an antagonistic coevolution going on where males are selected for higher amounts of toxins or other fitness increasing strategies and females for resistance to these harmful traits (Timmermeyer, 2008).

1.3 Evolution of androdioecy

Androdioecy is found when otherwise hermaphroditic species facultatively also produce males. Although androdioecy appears very difficult to evolve (Weeks et al., 2006a), there are a handful of cases. Androdioecy has been reported in several species of nematode, 2 orders of branchiopod crustacean and 17 barnacle species across 7 families (cirriped crustaceans) One controversial question is how often androdioecy arose within these groups (Weeks et al., 2006b)?. There is also one vertebrate species, an Osteichthyes fish, *Kryptolebias marmoratus*, which is the only vertebrate known to self-fertilize and also produce functional males.

The switch from gonochorism to hermaphroditism could occur when population densities are low enough that finding mates is difficult (Baker, 1955). At low population densities one would expect that hermaphrodites make only a small investment in male function because there is virtually no sperm competition such that small numbers of sperm are sufficient to fertilize the own eggs, in the case of self-fertility of ensure paternity upon cross-fertilization. This low investment of hermaphrodites in sperm production could later favor the evolution of androdioecy because males must have at least twice the male gametes fecundity of hermaphrodites in order to successfully invade a hermaphroditic population (Charlesworth, 1993).

Crossfertilizing hermaphrodites have all the advantages of sexual reproduction but also the disadvantages that only 50% of their genes go to the next generation. Deleterious genes cannot be easily purged from the population. On the other hand inbreeding depression is no problem. Selffertilizing hermaphrodites encounter, when they newly arise, the problem of inbreeding depression. On the other hand new mutations will be homozygozed immediately and deleterious mutations will be purged from the population. If inbreeding depression is low, selfing should be an advantage, because a high percentage of the genes are propagated to the next generation. With low number of deleterious mutations, homozygotes are produced and are purged from the population e.g. (Crnokrak & Barrett, 2002). If inbreeding depression is high, outcrossing should be an advantage, because there are more heterozygotes produced, which compensate for the loss of percentage of genes of the hermaphrodite which are not passed on. (e.g. (Stewart & Phillips, 2002).

Androdioecy could have evolved either by the invasion of a previously hermaphroditic species by males or by the replacement of the females with hermaphrodites in a previously gonochoristic species. It is quite difficult to imagine that a male arose in a purely selffertilizing hermaphroditic population – the whole genital apparatus would have to be newly invented. An androdioecious system with self-fertilizing hermaphrodites and males. like in *Caenorhabdtits elegans*, most probably descended from a male/ female system.

Because *Caenorhabditis elegans* has two alternative ways of reproduction, self-fertilization and outcrossing, it is an ideal system to study the advantages and disadvantages of the two modes of reproduction. As outlined above, *Caenorhabditis elegans* androdioecy is most probably derived from dioecy

(Kiontke et al., 2004). This hypothesis is further supported by the fact that its closest relatives are gonochoristic or androdioecious with selffertilizing hermaphrodites (Kiontke et al., 2004). An interesting question different from how androdioecy arose in Caenorhabditis elegans, is why it did not evolve further to pure hermaphroditism?. Inbreeding depression is low in Caenorhabditis elegans (Dolgin et al., 2007) and one could speculate that increasing levels of self-fertilization might ultimately lead to a situation where males are unable to produce offspring because all the hermaphrodites selfed exclusively. It is an ongoing discussion whether males are selectively maintained in natural populations or if they are evolutionary relics (Chasnov & Chow, 2002; Cutter et al., 2003; Stewart & Phillips, 2002).

Since I would like to contribute to answering this question I decided to study the role of males in C. elegans.

1.4 Nematode – basic morphology

The Nematoda are the phylum with the highest number of individuals, and the only place where one cannot find them is in the air (Lambshead, 1993). Body size can vary from a fraction of a mm to several meters and their body is covered by a cuticle. The body has no internal segmentation, and beneath the cuticle the wall is composed of an epidermis (hypodermis) and a single layer of muscle cells. The nematodes have a pseudocoelom, whereby tissue derived from the mesoderm only partly lines the fluid filled body cavity. In nematodes, as with rotifers (Phylum Rotifera), the body cavity is lined on the inside by endoderm and on the outside by mesoderm (Towle, 1989). Most nematodes exhibit sexual dimorphism. The male reproductive system opens to the exterior via the cloaca situated at posterior end of the body, while the female system opens via a pore called the vulva which is found in the ventral body wall (Wood, 1988).

1.5 Phylogenetic position of *Caenorhabditis elegans*

It is still debated where to place the nematoda within the metazoa. In one, currently widely accepted opinion, nematodes are united with Kinorhyncha, Priapulida, Nematomorpha, Tardigrada, Onychophora and Arthropoda into a clade of moulting animals, the Ecdysozoa (Aguinaldo *et al.*, 1997). In general Protostoma are to be split into Ecdysozoa and Lophotrochozoa. The latter include the Lophophorata with Brachiopoda, Ectoprocta, Phoronida and Entoprocta and the Trochozoa with Mollusca, Annelida, Nemertea, Sipunculida and Echiura (Nielson, 1995).

Besides evidence for a last common ancestor from the SSU rDNA (small subunit ribosomal DNA), also some morphological features argue for the monophyly within the Ecdysozoa (Aguinaldo *et al.*, 1997). Details of cuticle structure, molting sequence and hormones in arthropods and nematodes suggest homology, but there is a lack sufficient data on the minor 'ecdysozoan' phyla to rule out convergence (Schmidt-Rhaesa, 1998).

Alternatively, on the morphological level, the traditional Articulata- concept has more advantages, because the Arthropoda bauplan, in particular segmentation, can be explained convincingly from the Annelid ancestor. Traditionally, zoologists have regarded molluscs and annelids as the closest relatives of arthropods. Uniquely among protostomes, arthropods and annelids share the characteristic of adding segments from a posterior growth zone during ontogeny. Arthropods and annelids share with molluscs the additional combination of coelomic cavities with metanephridia, which also function as gonoducts, and a haemal system. (Meglitsch, 1991). The Ecdysozoa – concept cannot hold any similar explanation.

Caenorhabditis elegans belongs to the phylum nematoda,

Class Secernentea

Subclass Chromadoria

Order Rhabditida (polyphyletic group Blaxter 1998, Holterman 2006)

Suborder Rhabditina

Superfamily Rhabditoidea

Family Rhabditidae

(Chabaud, 1974; Lee, 2002).

1.6 Common aspects – lifestyle of *Caenorhabditis elegans*

Caenorhabditis elegans is an important model system for biological research. It has a short life cycle, compact genome, stereotypical development and it is easy to propagate in the lab and has a small size.

Caenorhabditis elegans has an androdioecious reproductive system, meaning it propagates either by self-ferilization or via cross-breeding with a male. During late larval development the hermaphrodites produce a limited number of sperm before they switch to oogenesis. The sperm is stored in the spermatheca until it is used for fertilization of the oocytes. In *Caenorhabditis elegans* the gender is determined over an XX/XO system. Females are XX and males XO. There is a low occurrence of males, that arise spontaneously as a result of X-chromosome non-disjunction during meiosis. Except for these males that make up around 0.2% in the standard laboratory strain (N2), all of the brood of an unmated hermaphrodite are hermaphroditic. Males can mate with hermaphrodites and give rise to 50% males in the cross-progeny. Cross-fertilization is not possible among hermaphrodites (Hope, 1999; Wood, 1988). The male sperm is also stored in the spermatheca and it is usually larger than the hermaphroditic sperm making it competitively advantageous (LaMunyon & Ward, 1998; LaMunyon & Ward, 1999).

On the one hand, there exists a large knowledge of *Caenorhabditis elegans* as a model experimental system. On the other hand there is minimal information available about its natural environment and ecology (Denver *et al.*, 2003). Dauer inducing conditions are a consistent selective pressure in natural populations as an overwhelming proportion of soil natural isolates are found in the dauer stage (Barriere & Felix, 2005). The dauer stage is an alternative 3rd larval stage formed under conditions of overcrowding, limited food, or high temperature (Cassada & and Russel, 1975). Until conditions improve they can remain in diapause for months (Klass & and Hirsch, 1976).

Caenorhabditis elegans has been found associated with diverse invertebrates (Sudhaus & Kiontke, 1996) including snails (genera *Helix*, *Oxychilus*, and *Pomatias elegans*), isopods (*Oniscusasellus*) and a *Glomeris* myriapod. This suggests that, unlike some other nematodes, *C. elegans* does not have a narrow host range. These associations are likely to contribute to the dispersal of *C. elegans* and may in addition be necromenic (Barriere & Felix, 2005). Many *Caenorhabditis elegans* strains from diverse geographical places have been isolated and characterized, beside the widely used N2 strain, isolated in Bristol, England (Hodgkin & Doniach, 1997). Natural *Caenorhabditis elegans* isolates vary in feeding behavior (de Bono & Bargmann, 1998), sperm size (LaMunyon & Ward, 2002), body length, fecundity, and other phenotypic characters (Hodgkin & Doniach, 1997).

It has been demonstrated that the lifespan of *Caenorhabditis elegans* in soil is much shorter than when grown under laboratory conditions. However, the shortened lifespan of worms in soil is more than sufficient for *Caenorhabditis elegans* to maintain viable populations (Van Voorhies et al., 2005). Even a single *Caenorhabditis elegans* can quickly produce a large population because of short generation time, high rate of progeny production, and self-fertilizing mode of reproduction (Venette & Ferris, 1998 ; Wood, 1988). Additional longevity adds little to its lifetime reproductive output, as *Caenorhabditis elegans* can produce almost all of its progeny within the first 2 days of its adult life (Byerly et al., 1976; Hirsh et al., 1976).

Sperm limits the number of progeny for unmated hermaphrodites. Increasing number of sperm would increase number of progeny produced. However, in a mutation that ends up with an increase in sperm production, the generation time is longer, due to a delay in the onset of oogenesis. One would therefore expect, that under natural conditions increased fecundity leads to a disadvantage. Evidence for this was also obtained in the lab and the optimal broodsize was estimated to lay within 250 and 350 per worm, similar to that of the N2 strain (Hodgkin & Barnes, 1991).

1.7 Why are there still males ?

Whether the continuing presence of males is adaptive or males are evolutionary relics is debatable and directly relevant to the long-standing and important problem as to why sexual reproduction and outcrossing is so widespread in nature (Chasnov & Chow, 2002; Cutter et al., 2003; Otto & Lenormand, 2002). Are *Caenorhabditis elegans* hermaphrodites descended from modified females capable of spermatogenesis that successfully invaded the ancestral dioecious species? If so, why are there still males present in the *Caenorhabditis. elegans* species (Chasnov & Chow, 2002)? Maintenance of males is poor in laboratory populations of *Caenorhabditis elegans* exhibiting little or no outcrossing (Cutter, 2005; Stewart & Phillips 2002; Teotonio et al., 2006). This low occurrence of males was the basis for the idea that *Caenorhabditis elegans* males are evolutionary relics without fitness benefit for the hermaphrodites (Chasnov & Chow, 2002). "Yet natural selection has maintained the complex genetic pathways of male development, suggesting that males may bestow an as-yet-unknown advantage to the species" (Prahlad *et al.*, 2003). To address these questions it is important to learn more about the various aspects of *C. elegans* reproduction, like outcrossing rates, mating behaviour and success and the dynamics of male appearance and disappearance in populations.

1.8 Outcrossing rates in natural populations

Sivasundar (2002) examined 20 microsatellite loci in a sample of 23 natural isolates of *Caenorhabditis elegans* from various parts of the world. The overall genetic variation was low and the genetic variation in local populations was almost as much as between populations all over the world. The reason for this cannot be easily determined. High levels of migration cannot explain the similarity among isolates, because one would expect a mixture of genotypes not a reduction. Also selection cannot be the reason for similarity because it would lead to rapid differentiation among local population at sites that are not under selection. Recent dispersal, perhaps by human association, followed by local population expansion could be the reason for populations being so similar (Phillips, 2006).

By measuring population genetic footmarks like linkage disequilibrium, heterozygosity or genetic diversity several authors have attempted to infer the out-crossing frequencies in natural populations (Barriere & Felix, 2005; Barriere & Felix, 2007; Cutter, 2006; Denver *et al.*, 2003; Sivasundar & Hey, 2002; Sivasundar & Hey, 2005). Males leave an appreciable genetic footprint in natural populations indicating that outcrossing does occur in wild populations. All authors agree that out-crossing is rare in natural populations with estimates of out-crossing rate varying between 10^{-5} and 0.02 except for Sivasundar, (2005) #79}, who estimated an out-crossing rate of 0.2. Nevertheless, even rare out-crossing events may be sufficient to reduce the mutational load and/or maintain sufficient genetic diversity required for rapid adaptation to fluctuating environments (Panell, 2002).

Ultimately all self-fertilizing populations may be at risk of mutation accumulation through inbreeding. Since the threat of extinction is slightly elevated under stressful conditions, in the face of environmental stress facultative outcrossing may enhance the adaptive response of high selfing population (Morran et al., 2009). Consistent with this idea, post dauer there are more males because males go into dauer more easily (Ailion & Thomas, 2000) and survive better (Morran and Phillips 2009). Additionally the CB4856 and JU440 strains exhibit dauer-induced elevated progeny number for both genders. This indicates altered mating dynamics in both hermaphrodites and males after going through dauer (Morran et al., 2009). Consequentionally, environmental stress is the key player in shifting the mating system from predominantly selfing to at least partially outcrossing. While the dauer stage has an adaptive value during stress, enhanced outcrossing occurs after the worms emerge from dauer and may help to adapt to new environments.

Laboratory evolution experiments have suggested that elevated mutation rates induced either by chemical mutagens (Manoel et al., 2007) or by a deficient DNA repair mechanism (Cutter, 2005) represent a selective force that favours higher male frequencies.

(Manoel et al., 2007) showed that induction of deleterious alleles diminishes the strength of selection against males in different genetic backgrounds. They carried out a male maintenance assay in the presence of the chemical mutagen ethane methyl sulfonate (EMS), and after 11 generations show a clear rise in male proportion compared with standard laboratory conditions. This phenomenon can be explained by the fact that hermaphrodites selfed and produced progeny which was homozygous for lethal mutations and led to the death of hermaphrodites.

Sex can be maintained when the deleterious mutation rate is sufficiently high. (Cutter, 2005) wanted to test the influence of mutation on the evolution of obligate outcrossing. He generated two strains of *Caenorhabditis elegans* with high and low mutations rate genetic backgrounds. Then he tracked the change in frequency of females, hermaphrodites and males over 21 generations. Experimental populations with elevated mutation rates experienced more outcrossing as well as greater retention of females.

Experimental populations of *Caenorhabditis elegans* were maintained either by outcrossing (sperm competition present) or by selfing (no sperm competition), and after 60 generations, significantly larger male sperm had evolved in the outcrossing population (LaMunyon & Ward, 2002). Under natural conditions there is no male/sperm competition and therefore sperm maybe remain small due to low levels of outcrossing.. On further analysis of the data LaMunyon (2006) did not find the elevated diversity they had expected for outcrossing populations. The diversity within selfing and outcrossing populations were the same. The authors postulate that this is an indication for a reduction of variability in the outcrossing population due to the fact that many genes hitchhike through sexual selection (LaMunyon *et al.*, 2006).

1.9 Mating behavior

Females of the genus *Caenorhabditis* produce a pheromone to attract males, and surprisingly the *Caenorhabditis* pheromone is not species-specific, with both *Caenorhabditis remanei* and *Caenorhabditis* sp. strain CB5161 females secreting a pheromone that attracts males of all four *Caenorhabditis* species tested, including *Caenorhabditis briggsae* and *Caenorhabditis elegans* (Chasnov et al., 2007). The *Caenorhabditis* pheromone is, however, sex-specific, with only females of *C. remanei* and *Caenorhabditis* sp. strain CB5161 secreting the pheromone and the pheromone attracting only males but not females or hermaphrodites (Chasnov et al., 2007). Furthermore, stage specificity exists in both female secretion and male detection of the female sex pheromone. Pheromone secretion by females and male chemotaxis toward females requires energy from both females and males, and natural selection should favor female secretion and male detection of sex pheromone only for those worms that are ready to mate. Chasnov et. al (2007) showed that secretion and detection of the pheromone clearly peaks in adulthood, as expected for a sex pheromone that facilitates immediate mating.

Additionaly (Srinivasan et al., 2008) showed the mating signal consists of a synergistic blend of at least three ascarosides in *Caenorhabditis. elegans*. In low concentrations they act as a potent male attractant and in high concentrations they induce dauer formation. When exposed to the mixture of three ascarosides *Caenorhabditis brenneri* and *Caenorhabditis remanei* males respond in a similar way to *Caenorhabditis elegans,* namely attractive. *Caenorhabditis briggsae* and *Caenorhabditis japonica* responded only weakly.

Hermaphroditic *Caenorhabditis briggsae*, gonochoristic *Caenorhabditis remanei* and *Caenorhabditis species 4* males produce a factor that mmobilizes females during copulation (reference). Also the vulva slit is stimulated to open, so that the male copulatory spicules can be inserted. *Caenorhabditis elegans* and *Caenorhabditis briggsae* hermaphrodites are not affected by this factor meaning that during evolution of internal self-fertilization, selection has been relaxed and hermaphrodites have lost the ability to respond to the male soporific-inducing factor (Garcia et al., 2007).

Mating behavior of the male can be split into several categories consisting of response, backward locomotion, turning and vulva localisation, spicule insertion and ejaculation. Response is defined as placing the tail flush on the hermaphrodite body, and moving backwards until the male tail reaches the head or tail whereby the male turns around in a sharp ventral coil. This backing behaviour continues until the male tail has contact with the vulva. The spicule is then inserted and sperm ejaculated into the hermaphrodite uterus. It is not mandatory to complete all sub-behavior for successful copulation (Wormbook, 2005).

At the end sex is maybe just so much fun.

2. Aim of the thesis

There are various hypothesis about the reasons for self-reproduction and outcrossing, respectively, but relatively few experimental studies have been carried out. *Caenorhabditis elegans* is a suitable organism investigating these questions, because it can self-fertilize and outcross. The main aim of my thesis is to discover whether *Caenorhabditis elegans* males have a function and if so investigate what might it be.

My thesis consists of three parts.

First I asked what causes the difference in male maintenance between two different strains (Wegewitz *et al.*, 2008).

Second: I tested if males might facilitate the spreading of genotypes into resident populations (Wegewitz *et al.*, 2009).

Third: I subjected populations of *Caenorhabditis elegans* to changing physical or biological stress and asked if this selection pressure leads to populations with higher male maintenance (Wegewitz *et al.*, 2009).

3. Results and Discussion

3.1 Experimental insight into the proximate causes of male persistence variation among two strains of the androdioecious *Caenorhabditis elegans* (Nematoda)

Viktoria Wegewitz, Hinrich Schulenburg and Adrian Streit

BMC Ecology 2008, **8**:12

Synopsis

Background

Different *C. elegans* strains maintain males at different rates, which at least in parts is determined genetically. The are a number of different factors that could contribute to this, including mating efficiency of the male or the hermaphrodite or self broodsize. I carried out systematic reciprocal crosses of *Caenorhabditis elegans* N2, a strain, which maintains males poorly and *Caenorhabditis elegans* CB4856, a strain with a high tendency to maintain males in the population. One male was placed together with fourteen new hermaphrodites every day until it ceased siring progeny. Three parameters were evaluated simultaneously: the number of successful copulations, the number of offspring per male, and the number of cross- and self-progeny per successfully mated hermaphrodite. For further investigation of the mating behavior I placed one male with fourteen hermaphrodites and scored body contact and spicule insertions at 14 observation time points over 9 hours. To determine if repeated mating increased the number of progeny a hermaphrodite can produce, I placed a hermaphrodite with 1, 3, 6, or 12 new young males every day until the hermaphrodite no longer produced progeny.

Main conclusion

CB4856 males had significantly more successful copulations, with around 17 compared to six for N2. CB4856 also had more offspring per male around 3000 compared to 1000 for N2. These two parameters were not significantly affected by the hermaphrodites used. The differences in progeny produced by a male was a consequence of its mating rate and not fertilization rate since the number of crossprogeny per mated hermaphrodite was not significantly different between the two strains. Consistent with this CB4856 males achieved a higher rate of body contacts and spicule insertions over time than N2 in the mating behavior assay. On the other hand the N2 hermaphrodites, independently of the type of male or the number of crossprogeny they had, produced significantly more self-progeny than CB4856. These factors together, the higher mating efficiency of CB4856 males as well as the higher self-progeny of N2 hermaphrodites, can explain the higher levels of male maintenance in CB4856.

Repeated mating of one hermaphrodite with one new male every day increased the total number of cross-progeny compared to the mating efficiency assay, especially for N2. Repeated mating with increasing number of males lead to a significant reduction of offspring number in both strains. This reduction was more pronounced in the N2 strain creating a loss of up to 53% compared to a maximum of 40% for CB4856. This can be a consequence of sexual conflict with an increase in male-male competition.

Author contributions

I did all the bench work and participated in the experimental design, the analysis of the data and the writing of the manuscript. HS participated in the experimental design and the data analysis. He did all the statistical analyses and he co-wrote the manuscript together with AS. AS participated in the experimental design and the data analysis. He coordinated the whole work and supervised the practical work. He co-wrote the manuscript together with HS. The contributions of HS and AS should be considered equal.

3.2. Do males facilitate the spread of novel phenotypes within populations of the androdioecious nematode *Caenorhabditis elegans*?

Viktoria Wegewitz, Hinrich Schulenburg and Adrian Streit

Journal of Nematology 41(3): 247-254. 2009

Synopsis

Background

Outcrossing occurs in wild populations, and even though it is a rare event it might be sufficient to reduce the mutational load and/or maintain sufficient genetic diversity required for rapid adaptation to fluctuating environments. Evolutionary laboratory experiments showed that elevated mutation rates act as a selective force towards higher male frequencies. In the experiment I asked, if particular phenotypes may spread and persist more easily if the invader is a male, a mated hermaphrodite or a virgin hermaphrodite. These invaders were marked with a transgene and added to stable populations of unmarked worms. Every three days the frequency of the transgenic phenotype was tracked and than the population was reduced again to 500 or 100 individuals.

In a second experiment I wondered if abiotic or biotic stress conditions also act as selective pressure in favor of higher male frequencies. These elevated male frequencies could possibly enhance the spread of novel advantageous phenotypes. To achieve variable genotypes where selection can act on I interbred N2 and CB4856, which showed very different rates of male maintenance. These hybrid populations were exposed to varying environmental conditions or standard laboratory conditions.

Conclusion

When the reporter gene was brought to the population by a male or a mated hermaphrodite, gfp positive worms reached higher frequencies than after addition of the marker through a non-mated hermaphrodite. These results illustrate that rare males in a hermaphroditic population cause an increase in frequency of their alleles. This is obviously also the case for genes involved in male formation and development. If the occasional boost of frequency of functional alleles of these genes caused by the sporadic males is large enough to offset the loss of functional alleles by mutational degradation and drift (which is expected to happen in hermaphrodites), this might be sufficient to maintain the genetic machinery for the production of males, even if there is no fitness advantage of outcrossing for hermaphrodites.

However, it might also be advantageous for a mother to produce males, because this should indirectly lead to an increase of the frequency of her alleles. This advantage is expected to be strongly enhanced if novel environmental conditions (i.e. new selective constraints) can be expected to favor new phenotypes, because such new phenotypes are more rapidly produced through outcrossing and recombination than a series of mutations.

Second experiment: No difference in male maintenance between the populations, which had been subjected to the selection of changing stress and the control treatments was found. Both, the control as well as the selection populations had adopted a high male maintenance, indistinguishable from CB4856 but significantly different from N2. During the experimental evolution experiment small numbers of males of both parental strains were added to prevent stochastic loss of genotypes. To exclude that the dominance of the CB4856-like phenotype was caused by the addition of these males I repeated the control experiment in a simplified form (chunking) without the periodic addition of males. Again, high male maintenance significantly different from N2 and undistinguishable from CB4856 was reached again in all heterogeneous populations. This indicates that higher maintenance of males or something that causes it indirectly might be favourable even under standard laboratory conditions. It still remains to be addressed in the future whether high male maintenance does allow the population to adapt to new selective constraints more rapidly, because of the more generation of novel genotypes through outcrossing and recombination.

Author contributions

I did all the bench work and participated in the experimental design, the analysis of the data and the writing of the manuscript. HS participated in the experimental design and the data analysis. He did all the statistical analyses and he co-wrote the manuscript together with AS. AS participated in the experimental design and the data analysis. He coordinated the whole work and supervised the practical work. He co-wrote the manuscript together with HS. The contributions of HS and AS should be considered equal.

Bibliography

AGUINALDO, A. M., TURBEVILLE, J. M., LINFORD, L. S., RIVERA, M. C., GAREY, J. R., RAFF, R. A. and LAKE, J. A. (1997). Evidence for a clade of nematodes, arthropods and other moulting animals. *Nature*, **387**, 489-493.

AILION, M. and THOMAS, J. H. (2000). Dauer formation induced by high temperatures in Caenorhabditis elegans. *Genetics*, **156**, 1047-1067.

BAKER, H. G. (1955). Self compatibility and establishment after "long-distance" dispersal. *Evolution*, **9**, 347-349.

BARRIERE, A. and FELIX, M. A. (2005). Natural variation and population genetics of Caenorhabditis elegans. *WormBook*, 1-19.

BARRIERE, A. and FELIX, M. A. (2007). Temporal dynamics and linkage disequilibrium in natural Caenorhabditis elegans populations. *Genetics*, **176**, 999-1011.

BATEMAN, A. J. (1948). Intra-sexual selection in *Drosophila*. *Heredity*, **2**, 349-368.

BELL, G. (1982). *Masterpiece of Nature: The evolution of sexuality,* University of California Press, Berkeley.

BONHOEFFER, S., CHAPPEY, C., PARKIN, N. T., WHITCOMB, J. M. and PETROPOULOS, C. J. (2004). Evidence for positive epistasis in HIV-1. *Science*, **306**, 1547-1550.

BYERLY, L., CASSADA, R. C. and RUSSELL, R. L. (1976). The life cycle of the nematode Caenorhabditis elegans. I. Wild-type growth and reproduction. *Dev Biol*, **51**, 23-33.

CARROLL, L. (1865). *Alice in Wonderland,* Macmillan.

CASSADA, R. C. and AND RUSSEL, R. L. (1975). The dauer larva, a postembryonic developmental variant of the nematode *Caenorhabditis elegans*. *Dev. Biol.*, **46**, 326-342.

CHABAUD, A. G. (1974). *Keys to the Nematode Parasites of Vertebrates,* CAB, Farnham Royal, U.K.

CHARLESWORTH, B. (1993). The evolution of sex and recombination in a varying environment. *J Hered*, **84**, 345-350.

CHASNOV, J. R. and CHOW, K. L. (2002). Why are there males in the hermaphroditic species Caenorhabditis elegans? *Genetics*, **160**, 983-994.

CHASNOV, J. R., SO, W. K., CHAN, C. M. and CHOW, K. L. (2007). The species, sex, and stage specificity of a Caenorhabditis sex pheromone. *Proc Natl Acad Sci U S A*, **104**, 6730-6735.

COX, M. M. (2001). Historical overview: searching for replication help in all of the rec places. *Proc. Natl. Acad. Sci. U.S.A.*, **98**, 8173-8180.

CRNOKRAK, P. and BARRETT, S. C. (2002). Perspective: purging the genetic load: a review of the experimental evidence. *Evolution*, **56**, 2347-2358.

CUTTER, A. D. (2005). Mutation and the experimental evolution of outcrossing in Caenorhabditis elegans. *J Evol Biol*, **18**, 27-34.

CUTTER, A. D. (2006). Nucleotide polymorphism and linkage disequilibrium in wild populations of the partial selfer Caenorhabditis elegans. *Genetics,* **172,** 171-184.

CUTTER, A. D., AVILES, L. and WARD, S. (2003). The proximate determinants of sex ratio in C. elegans populations. *Genet Res,* **81,** 91-102.

DE BONO, M. A. and BARGMANN, C. I. (1998). Natural variation in neuropeptide Y receptor homolog modifies social behavior and food response in C. elegans. *Cell,* **94,** 679-689.

DENVER, D. R., MORRIS, K. and THOMAS, W. K. (2003). Phylogenetics in Caenorhabditis elegans: an analysis of divergence and outcrossing. *Mol Biol Evol,* **20,** 393-400.

DOLGIN, E. S., CHARLESWORTH, B., BAIRD, S. E. and CUTTER, A. D. (2007). Inbreeding and outbreeding depression in Caenorhabditis nematodes. *Evolution,* **61,** 1339-1352.

FISHER, R. A. (1930). *The genetic theory of natural selection,* Clarenon Press, Oxford, UK.

GARCIA, L. R., LEBOEUF, B. and KOO, P. (2007). Diversity in mating behavior of hermaphroditic and male-female Caenorhabditis nematodes. *Genetics,* **175,** 1761-1771.

GAVRILETS, S., ARNQVIST, G. and FRIBERG, U. (2001). The evolution of female mate choice by sexual conflict. *Proc Biol Sci,* **268,** 531-539.

GÖLDENBOOG, C. (2006). *Wozu Sex,* Deutsche-Verlags-Anstalt, München.

HAMILTON, W. D., AXELROD, R. and TANESE, R. (1990). Sexual reproduction as an adaptation to resist parasites (a review). *Proc Natl Acad Sci U S A,* **87,** 3566-3573.

HIRSH, D., OPPENHEIM, D. and KLASS, M. (1976). Development of the reproductive system of Caenorhabditis elegans. *Dev Biol,* **49,** 200-219.

HODGKIN, J., AND and DONIACH, T. (1997). Natural variation and copulatory plug formation in Caenorhabditis elegans. *Genetics,* **146,** 149-164.

HODGKIN, J. and BARNES, T. M. (1991). More is not better: brood size and population growth in a self-fertilizing nematode. *Proc Biol Sci,* **246,** 19-24.

HOPE, I. A. (1999). Background on Caenorhabditis elegans. In C. elegans a practical approach, Oxford: Oxford University Press.

JOHNSTONE, R. A. A. and KELLER, L. (2000). How males can gain by harming their mates: sexual conflict, seminal toxins, and the cost of mating. *Am. Nat.,* **156,** 368-377.

KIONTKE, K., GAVIN, N. P., RAYNES, Y., ROEHRIG, C., PIANO, F. and FITCH, D. H. (2004). Caenorhabditis phylogeny predicts convergence of hermaphroditism and extensive intron loss. *Proc Natl Acad Sci U S A,* **101,** 9003-9008.

KLASS, M. and AND HIRSCH, D. (1976). Non-ageing developmental variant of Caenorhabditis elegans. *Nature,* **260,** 523-525.

KONDRASHOV, A. S. (1988). Deleterious mutations and the evolution of sexual reproduction. *Nature,* **336,** 435-440.

LAMBSHEAD, J. (1993). Recent development in marine benthic biodiversity research. *Oceanis,* **19,** 5-24.

LAMUNYON, C. W., BOUBAN, O. and CUTTER, A. D. (2006). Postcopulatory sexual selection reduces genetic diversity in experimental populations of Caenorhabditis elegans. *J Hered*, **98**, 67-72.

LAMUNYON, C. W. and WARD, S. (1998). Larger sperm outcompete smaller sperm in the nematode Caenorhabditis elegans. *Proc Biol Sci*, **265**, 1997-2002.

LAMUNYON, C. W. and WARD, S. (1999). Evolution of sperm size in nematodes: sperm competition favours larger sperm. *Proc Biol Sci*, **266**, 263-267.

LAMUNYON, C. W. A. and WARD, S. (2002). Evolution of larger sperm in response to experimentally increased sperm competition in Caenorhabditis elegans. *Proc. R. Soc. Lond. B Biol. Sci.*, **269**, 1125-1128.

LEE, D. L. (2002). *The biology of nematodes,* Taylor & Francis.

LESSELLS, C. M. (2005). Why are males bad for females? Models for the evolution of damaging male mating behavior. *Am Nat*, **165 Suppl 5**, S46-63.

MANOEL, D., CARVALHO, S., PHILLIPS, P. C. and TEOTONIO, H. (2007). Selection against males in *Caenorhabditis elegans* under two mutational treatments. *Proc. R. Soc. B*, **274**, 417-424.

MAYNARD SMITH, J. (1978). *The Evolution of Sex,* Cambridge University Press.

MAYNARD-SMITH, J. (1978). *The evolution of sex,* Cambridge University Press, Cambridge, UK.

MEGLITSCH, P. A., AND SCHRAM F.R. (1991). *Invertebrate Zoology,* Oxford Univ. Press, New York.

MICHIELS, N. K. (1998). Mating conflicts and sperm competition in simultaneous hermaphrodites. In *Sperm competition and sexual selection* (eds. Birkhead, T. R. a. & Moller, A. P.), pp. 219-254. Academic press, London.

MORRAN, L. T., CAPPY, B. J., ANDERSON, J. L. and PHILLIPS, P. C. (2009). Sexual Partners for the Stressed: Facultative Outcrossing in the Self-Fertilizing Nematode *C. Elegans. Evolution*.

MORROW, E. H., ARNQVIST, G. and PITNICK, S. (2003). Adaptation versus pleiotropy: Why do males harm their mates? . *Behav. Ecol.*, **14**, 802-806.

MULLER, H. J. (1964). The Relation of Recombination to Mutational Advance *Mutat Res*, **106**, 2-9.

NIELSON, C. (1995). *Animal Evolution. Interrelationships of the living phyla.*

OTTO, S. P. and LENORMAND, T. (2002). Resolving the paradox of sex and recombination. *Nat Rev Genet*, **3**, 252-261.

PANELL, J. (2002). The evolution and maintenance of androdioecy. *Annu Rev Ecol Syst*, **33**, 397-425.

PARKER, G. A. (1970). Sperm competition and its evolutionary consequences in the insects. *Biol. Rev.*, **45**, 525-567.

PHILLIPS, P. C. (2006). One perfect worm. *Trends Genet*, **22**, 405-407.

PRAHLAD, V., PILGRIM, D. and GOODWIN, E. B. (2003). Roles for mating and environment in C. elegans sex determination. *Science*, **302**, 1046-1049.

SCHMIDT-RHAESA, A. (1998). Muscular ultrastructure in Nectonema munidae and Gordius aquaticus (Nematomorpha). *Invert. Biol.,* **117,** 37-44.

SIVASUNDAR, A. and HEY, J. (2002). Population Genetics of *Caenorhabditis elegans*: The Paradox of Low Polymophism in a Widespread Species. *Genetics,* **163,** 147-157.

SIVASUNDAR, A. and HEY, J. (2005). Sampling from natural populations with RNAI reveals high outcrossing and population structure in Caenorhabditis elegans. *Curr Biol,* **15,** 1598-1602.

SRINIVASAN, J., KAPLAN, F., AJREDINI, R., ZACHARIAH, C., ALBORN, H. T., TEAL, P. E., MALIK, R. U., EDISON, A. S., STERNBERG, P. W. and SCHROEDER, F. C. (2008). A blend of small molecules regulates both mating and development in Caenorhabditis elegans. *Nature,* **454,** 1115-1118.

STEWART, A. D. and PHILLIPS, P. C. (2002). Selection and maintenance of androdioecy in Caenorhabditis elegans. *Genetics,* **160,** 975-982.

STOCKLEY, P., SEARLE, J. B., MACDONALD D.W. and JONES, C. S. (1995). Correlates of reproductive success within alternative mating tactics of the common shrew. *Behavioral Ecology,* **7,** 334-340.

SUDHAUS, W. and KIONTKE, K. (1996). Phylogeny of Rhabditis subgenus Caenorhabditis (Rhabditidae, Nematoda). *J. Zoo.Syst. Evol. Research* **34,** 217-233.

TEOTONIO, H., MANOEL, D. and PHILLIPS, P. C. (2006). Genetic variation for outcrossing in *Caenorhabditis elegans. Evolution,* **60,** 1300-1305.

TIMMERMEYER, N. (2008). Sexual conflict and mating behavior in nematodes. Diploma thesis.

TOWLE, A. (1989). *Modern Biology,* Holt, Rinehart and Winston, Austin, TX.

VAN VOORHIES, W. A., FUCHS, J. and THOMAS, S. (2005). The longevity of Caenorhabditis elegans in soil. *Biol Lett,* **1,** 247-249.

VENETTE, R. and FERRIS, H. (1998). Influence of bacterial type and density of popualtion growth of bacterial-feeding nematodes. *Soil Biol. Biochem.,* **30,** 949-960.

WEEKS, S. C., BENVENUTO, C. and REED, S. K. (2006a). When males and hermaphrodites coexist: a review of androdioecy in animals. *Integrative and comparative Biology,* **46,** 449-464.

WEEKS, S. C., SANDERSON, T. F., REED, S. K., ZOFKOVA, M., KNOTT, B., BALARAMAN, U., PEREIRA, G., SENYO, D. M. and HOEH, W. R (2006b). Ancient androdioecy in the freshwater crustacean Eulimnadia. *Proc Biol Sci,* **273,** 725-734.

WEGEWITZ, V., SCHULENBURG, H. and STREIT, A. (2008). Experimental insight into the proximate causes of male persistence variation among two strains of the androdioecious Caenorhabditis elegans (Nematoda). *BMC Ecol,* **8,** 12.

WEHNER, G. A. and GEHRING, W. (2007). *Zoologie,* Georg Thieme Verlag KG.

WEISMANN, A. (1889). *Essay on heredity and kindred biological subjects,* Oxford Univ. Press, Oxford, UK.

WILLIAMS, G. C. (1975). *Sex and Evolution,* Princeton University Press.

WOOD, W. B. (1988). Introduction to *C. elegans* biology. In *The nematode Caenorhabditis elegans* (ed. W.B., W.), pp. 1-16. Cold Spring Harbor Laboratory Press.
WORMBOOK (2005). Wormbook: The Online Review of C. elegans Biology. Pasadena (CA).

Summary

At first sight sexual reproduction seems to have only disadvantages in comparison to asexual reproduction. Two mates have to find each other and then they only give half of their chromosomes to the next generation. Sexual reproduction includes double the costs, because essentially unproductive males have to be produced to fertilize the females. In asexual reproduction only females are produced, all of which can directly propagate again. But there are also advantages of sexual reproduction, which probably act together. Sexual reproduction increases the variability between siblings. Further, sex enables previously separated advantageous alleles to recombine into one genome. Recombination also leads to the separation of positive genes from disadvantageous genetic backgrounds, thereby facilitating their dispersal.

Caenorhabditis elegans exists predominantly as a self-fertilizing hermaphrodite. This represents an extreme case of inbreeding, which in many respects comes close to asexual reproduction. However, there are a few spontaneous males produced. They are fully functional and, upon mating with a hermaphrodite, sire 50% males and 50% hermaphrodites. Since *Caenorhabditis elegans* can fertilize itself as well as mate with a male, it is an ideal system, to investigate under which conditions males are advantageous.

It was known that males disappear quickly from laboratory populations of some strains (e.g. N2) while other strains can maintain a certain level of males over a longer time period of time (e.g. CB4856). In the first part of my work I quantified this effect for several isolates of *Caenorhabditis elegans* and I could show that a higher mating efficiency of CB4856 males and a higher self-fertilization rate of N2 hermaphrodites contribute to the difference in male maintenance between these two strains.

In the second part it was examined, how the genes of a single immigrating individual can invade and spread within a given population. Individuals, genetically marked with a green fluorescent protein (gfp) marker gene (invaders), were added to "stable" populations of different sizes and the dynamics of the *gfp* reporter gene was monitored over time. As invaders males, cross-fertilized hermaphrodites and virgin hermaphrodites were tested. Addition of fertilized hermaphrodites, which produce nearly 50% males, achieved the highest frequency of the marker gene. If this occasional boost of males is large enough to offset the loss of functional alleles by mutational degradation and drift. It might be sufficient to maintain the genetic machinery for the production of males. Outcrossing might also be advantageous for a hermaphrodite because it should indirectly increase the frequency of her alleles. In addition, through recombination a genetically more diverse progeny is created. This in turn would increase the chance that among the offspring there a some, which are well-adapted to a novel environment.

In the third part I investigated whether cross-fertilization is advantageous under changing environmental conditions,. I generated genetically heterogeneous starting populations by interbreeding N2 and CB4856 and subjected them to either a selection regime of changing environmental conditions or standard laboratory conditions. Contrary to the expectation, I found no difference between the two treatments. Interestingly, at the end of the experiment all populations showed a high tendency of male maintenance, similar to CB4856. This indicates that higher maintenance of males or something that causes it indirectly might be favourable even under standard laboratory conditions.

Zusammenfassung der Doktorarbeit

Sexuelle Reproduktion hat doppelte Kosten, denn es müssen Männchen gemacht werden, die die Weibchen befruchten. Bei der asexuellen Reproduktion werden nur Weibchen und damit doppelt soviele Nachkommen erzeugt, welche sich direkt wieder fortpflanzen können. Aber es gibt auch Eigenschaften von sexueller Fortpflanzung, die wahrscheinlich vorteilhaft sind und wohl zusammen wirken. Sexuelle Reproduktion erhöht die Variabilität zwischen den Nachkommen. Zum anderen wenn zwei Genome durch Sex gemischt werden, können Allele, welche von Vorteil sind innerhalb weniger Generationen durch Rekombination kombiniert werden. Die Verbreitung von Vorteil habenden Gene könnte erleichtert werden, in dem sie aus ihrem genetischen Hintergrund herausgelöst werden können. Im Idealfall kommen auf ein Chromosom nur die guten Gene und auf ein anderes nur die schlechten. Der Nematode *Caenorhabditis elegans* kommt vor allem als sich selbst befruchtender Hermaphrodit vor. Dies stellt einen extremen Fall von Inzucht dar und kommt in vielerlei Hinsicht einer asexuellen Fortpflanzung nahe. Allerdings werden spontan auch einige wenige Männchen gebildet. Diese sind voll funktionsfähig und produzieren, wenn sie sich mit einem Hermaphroditen paaren, 50% Männchen und 50% Hermaphroditen. Da sich *Caenorhabditis elegans* sowohl durch Selbstbefruchtung wie auch durch Kreuzbefruchtung mit Männchen fortpflanzen kann, ist es ein ideales System, um zu untersuchen, unter welchen Bedingungen Männchen von Vorteil sein könnten. Es war bekannt, dass bei einigen Laborstämmen von *Caenorhabditis elegans* Männchen aus den Kulturen verloren gehen (z. B. N2), während andere Stämme einen gewissen Anteil an Männchen über längere Zeit aufrecht erhalten (z.B. CB4856). Im ersten Teil meiner Arbeit konnte ich zeigen, dass ein höherer Paarungserfolg von CB4856 Männchen und eine höhere Selbstbefruchtungsrate von N2 Hermaphroditen zu diesem Unterschied beitragen.

Im zweiten Teil wurde untersucht, in wie weit einzelne Individuen, die in eine Population einwandern, ihre Gene in diese Population einbringen können. Genetisch markierte Männchen, gepaarte Hermaphroditen oder jungfräuliche Hermaphroditen wurden zu einer Population zugeführt. Es stellte sich heraus, dass Zugabe von gepaarten Hermaphroditen, welche fast 50% Männchen produzieren, zu den höchsten Allelfrequenzen des Markergens führte. Vielleicht reicht der gelegentliche Anstieg an Männchen aus, um das Verlorengehen funktionierender Allele durch Mutation und Drift auszugleichen. Auf der anderen Seite ist es vielleicht auch für einen kreuzbefruchteten Hermaphroditen eine hohe Männchenrate von Vorteil, weil es indirekt die Frequenz der eigenen Allele erhöht. Dieser Vorteil ist besonders stark, wenn neue Umweltbedingungen zu erwarten sind, an denen sich die neuen Phänotypen anpassen können.

Im dritten Teil sollte eine theoretische Voraussage, welche besagt dass unter sich ändernden Umweltbedingungen, eine höhere Kreuzbefruchtungsrate von Vorteil sei, getestet werden. Um eine genetisch variable Ausgangspopulation zu erzeugen wurden N2 und CB4856 gekreuzt. Die eine Hälfte der Populationen wurden sich ändernden Umweltbedingungen ausgesetzt die andere Hälfte wurde unter Standartbedingungen gehalten. Entgegen der Voraussage, wiesen am Schluss alle Populationen, und nicht nur diejenigen welche variablen Umweltbedingungen ausgesetzt waren, eine hohe Tendenz zum Erhalten von Männchen auf, ähnlich wie CB4856. Das impliziert, dass eine höhere Männchenerhaltungsrate oder etwas, dass das indirekt verursacht, von Vorteil ist auch unter Standardbedingungen.

BMC Ecology

Research article

Open Access

Experimental insight into the proximate causes of male persistence variation among two strains of the androdioecious *Caenorhabditis elegans* (Nematoda)

Viktoria Wegewitz[1], Hinrich Schulenburg[2] and Adrian Streit*[1]

Address: [1]Department of Evolutionary Biology, Max Planck Institute for Developmental Biology, Tübingen, Germany and [2]Institute of Zoology, University of Tübingen, Tübingen, Germany

Email: Viktoria Wegewitz - viktoria.wegewitz@tuebingen.mpg.de; Hinrich Schulenburg - hinrich.schulenburg@uni-tuebingen.de; Adrian Streit* - adrian.streit@tuebingen.mpg.de

* Corresponding author

Published: 13 July 2008

BMC Ecology 2008, **8**:12 doi:10.1186/1472-6785-8-12

Received: 7 December 2007
Accepted: 13 July 2008

This article is available from: http://www.biomedcentral.com/1472-6785/8/12

© 2008 Wegewitz et al; licensee BioMed Central Ltd.
This is an Open Access article distributed under the terms of the Creative Commons Attribution License (http://creativecommons.org/licenses/by/2.0), which permits unrestricted use, distribution, and reproduction in any medium, provided the original work is properly cited.

Abstract

Background: In the androdioecious nematode *Caenorhabditis elegans* virtually all progeny produced by hermaphrodite self-fertilization is hermaphrodite while 50% of the progeny that results from cross-fertilization by a male is male. In the standard laboratory wild type strain N2 males disappear rapidly from populations. This is not the case in some other wild type isolates of *C. elegans*, among them the Hawaiian strain CB4856.

Results: We determined the kinetics of the loss of males over time for multiple population sizes and wild isolates and found significant differences. We performed systematic inter- and intra-strain crosses with N2 and CB4856 and show that the males and the hermaphrodites contribute to the difference in male maintenance between these two strains. In particular, CB4856 males obtained a higher number of successful copulations than N2 males and sired correspondingly more cross-progeny. On the other hand, N2 hermaphrodites produced a higher number of self-progeny, both when singly mated and when not mated.

Conclusion: These two differences have the potential to explain the observed variation in male persistence, since they should lead to a predominance of self-progeny (and thus hermaphrodites) in N2 and, at the same time, a high proportion of cross-progeny (and thus the presence of males as well as hermaphrodites) in CB4856.

Background

The nematode *Caenorhabditis elegans* is a facultative hermaphrodite that reproduces either by virtue of self-fertilization or cross-breeding with a male (androdioecious reproductive system). The hermaphrodites are somatically female but produce a limited number of sperm during their late larval development before switching to the production of eggs. The sperm is stored in the spermatheca and can be used to fertilize the newly formed eggs. Except for a very few males (around 0.2% in the standard laboratory strain N2) that arise spontaneously as the result of X chromosome non-disjunction, the entire self-progeny is hermaphroditic. Males can mate with hermaphrodites and give rise to 50% males in the cross-progeny. Cross-fertilization is not possible among hermaphrodites [1,2]. Male derived sperm is also stored in the spermatheca where it competes with the hermaphrodite's own sperm for the fertilization of the oocytes. Male sperm is usually larger and therefore has a competitive advantage over the hermaphrodite's sperm [3,4].

If hermaphrodites can reproduce by self-fertilization, males are superfluous. In fact, they could even represent a burden, decreasing individual fitness, in analogy with the two-fold cost of males in theories on the evolution of sex [5,6]. Therefore, the persistence of males represents an important puzzle for our understanding of *C. elegans* biology. Its explanation is expected to advance a more general insight into the evolution of androdioecy. To date, the function of *C. elegans* males has been addressed using two main approaches: (i) experimental evolution in the laboratory, and (ii) analysis of (male-dependent) outcrossing rates in wild populations.

The experimental evolution of laboratory populations of the standard strain N2 (as well as mutants derived from this strain) uniformly demonstrated that initial male frequencies of either 50% or 33% rapidly and steeply decline to less than 10% within ten to 15 generations [7-11]. Furthermore, compared with the dioecious species *Caenorhabditis remanei*, mating behaviour was severely compromised in the N2 strain, i.e males often fail to find hermaphrodite mates, possibly due to limited production and/or degeneracy of the hermaphrodite's sex pheromone [[7,12] but see, [13,14]]. Taken together, these results suggested that males represent evolutionary relics without any particular function, which are only still present, because of a relatively recent switch to hermaphroditism and selfing in the lineage leading to *C. elegans* [7]. Interestingly, however, other natural isolates show clear differences to N2. The spontaneous production of males is a pre-requisite for male maintenance and it reaches values of more than 3% of the total offspring in some isolates – clearly more than N2 with a value of less than 0.5% [11,15]. Similarly, males are able to persist in populations of some strains, e.g. the Hawaiian strain CB4856 and the Oregon strain PX174 [11]. This effect seems to be enhanced in these two strains (but not others) if worm populations are subjected to fluctuating environmental conditions like variable exposure to different mutagens [16]. Similarly, populations with deficient DNA repair and thus increased mutations rates also maintain males at higher frequencies [10]. These results suggested that males are beneficial to ensure frequent outcrossing, which is favored under variable environmental conditions and/or high deleterious mutation rates [17-22].

An alternative albeit indirect route to assess the function of males is to infer outcrossing rates in natural populations. Several recent studies analysed new *C. elegans* isolates from different parts of the world using a variety of molecular markers such as microsatellites, AFLPs, or DNA sequence polymorphisms. They unanimously demonstrate that outcrossing does occur, but that it is usually extremely rare [23-27]. The only exception is an inferred outcrossing rate of 0.2 [28], whereas all other studies suggest it to range in between 10^{-5} up to 0.02 [23-25]. Consequently, males leave a genetic footprint in natural populations. In consistency with the conclusions from experimental evolution in the lab, rare outcrossing may be sufficient to eliminate mutational load and/or maintain genetic diversity required for rapid adaptation to fluctuating environments [17,29,30].

In the first part of this publication we describe the decline of the proportion of males in eight different natural isolates under standard laboratory conditions. The fact that males are lost at very different rates even if the strains are maintained under the same conditions indicates that the difference is genetically determined and is therefore a putatively selectable trait. In the second part we evaluate the possible reasons for the difference in male persistence between the two common laboratory strains N2 and CB4856. Several behavioral and physiological factors could account for this difference, for example: i) the mating efficiency of the males, ii) the mating efficiency of the hermaphrodites (this includes the attractiveness of the hermaphrodites for males), iii) the competitive advantage of the male derived sperm, iv) the number of sperm transferred, v) the difference between the maximum number of progeny a hermaphrodite can produce with and without mating. In order to address these points, we performed a systematic analysis of intra- and inter-strain crosses between N2 and CB4856- to our knowledge for the very first time in this context. Our results indicate that CB4856 males are capable of mating successfully with more hermaphrodites than N2 males and that N2 hermaphrodites produced a higher number of (all-hermaphrodite) self-progeny even after mating. Both these effects result in a higher proportion of males in the next generation for CB4856 if compared with N2, thus potentially explaining male persistence in the former but not the latter strain.

Methods
C. elegans cultures

C. elegans was cultured as described in [31]. The preparation of NGM plates and *Escherichia. coli* strain OP50 food bacteria and M9 buffer is also described in [31].

Mating plates: 6 cm NGM plates were seeded with 30 μl of an *E. coli* (OP50) culture such that the plates contained a small round dot of bacteria in the center.

All experiments were done in an air-conditioned room at a temperature of 21 ± 1°C and 40% humidity. To minimize fluctuations of physical conditions plates were kept in boxes, randomized in piles that were placed evenly distributed within the boxes.

Strains used

N2: Standard laboratory wild type strain, isolated in Bristol, UK

CB4856: Standard polymorphic mapping strain, isolated in Hawaii.

AB1: isolate from Australia

JU258: isolate from Madeira

MY1, MY15, MY18, RC301: isolates from Germany

All strains are available from the *Caenorhabditis* Genetics Center at the University of Minnesota [32].

Male maintenance assay

This assay served to determine the persistence of males over time in different natural *C. elegans* isolates and in different population sizes. For each strain, we set up several crosses using a male:hermaphrodite ratio of 2:1. These crosses yielded populations with a gender ratio of approximately 1:1. For each strain, the populations were mixed four days after setting up the crosses and a defined number of individuals (population size) was randomly chosen and transferred onto NGM plates for the experiment (day 0 of the experiment). These experimental populations were all treated as follows: After three days adult males and hermaphrodites were counted (counting, see below). On the next day the population was reduced to the original population size and transferred to a new plate (transfer, see below). Three days later, adult males and hermaphrodites were counted again, followed by population size reduction and worm transfer one day later, as above. This whole procedure was repeated for a total of eight times (equivalent to 32 days). All male maintenance assays were done on 9 cm NGM plates, seeded with 1 ml of *E. coli* OP50 culture. Two sets of experiments were performed: i) two replica runs per strain and population size were done in parallel for strains N2, CB4856, AB1, JU258, MY1, MY15, MY18 and RC301 using population sizes of 75 and 150; and ii) five replica runs per strain and population size were done in parallel for N2 and CB4856 using population sizes of 40, 70, 100 and 150. In this context, one important objective was to evaluate the effect of different population sizes on male persistance. The exact numbers used (e.g. 75 versus 150) were chosen arbitrarily.

Counting

The plates were placed under a dissecting microscope and searched systematically always using the same search path with the help of a grid, which was positioned below the plates. The first 100 to 120 adults encountered were then used to determine the number of adult hermaphrodites and adult males. Note that this is equivalent to a random choice of individuals. Only hermaphrodites with developing embryos in the uterus were counted as adult hermaphrodites. This may have lead to a slight underestimation of the number of hermaphrodites.

Transfer

Worms were washed off plates with M9 buffer and counted without paying attention to the developmental stage or sex of the worms. The volume that was expected to contain the desired number of worms was transferred onto a new plate. Thus, the populations always consisted of mixed generations, so that the effective reproductive population was smaller than the actual number of animals.

Male mating efficiency assay

This assay served to evaluate the mating efficiency of males in terms of mated hermaphrodites, total offspring produced per male and also cross- as well as self-progeny produced per mated hermaphrodite. A single male was confronted with an excess of hermaphrodites, so that it could mate as often as possible. In a pilot experiment, 14 hermaphrodites were found to be sufficient to ensure that the male would never come close to mating with all of them. In fact, during the main experiment the highest number of mated hermaphrodites per plate was 9.

Mating plates were prepared four days prior to the experiment. One male and 14 young adult hermaphrodites were placed on mating plates. Every 24 hours the male was moved to a new mating plate with 14 young adult hermaphrodites until no more successful mating was observed. Hermaphrodites, which were exposed to males, were placed individually on NGM plates seeded with 200 μl *E. coli* OP50. The hermaphrodites were transferred to new plates every 24 h. After three days the progeny was counted or, if the number of plates was too high to be processed immediately, equal numbers of plates from both treatments were put at 4°C and counted within a few days. This step allowed us to do more replicas in parallel and have all plates scored by the same person, in order to avoid possible observer biases. We did not observe any lethality in response to the cooling step. A hermaphrodite was considered to be mated when more than one male was found among the progeny (successful mating event). The number of cross progeny per mated hermaphrodite was estimated as twice the number of male progeny. Since it was not feasible to do all crosses in parallel, we used an incomplete block design, where two different crosses were set up in parallel and different pairs of parallel crosses were assayed in four experimental runs: i) N2 males crossed with N2 and with CB4856 hermaphrodites; ii)

CB4856 males crossed with N2 and with CB4856 hermaphrodites; iii) N2 and CB4856 males crossed with N2 hermaphrodites; iv) N2 and CB4856 males crossed with CB4856 hermaphrodites.

Mating behavior assays
These two assays were used to characterize in more detail the time required by males until first contact with a hermaphrodite and first spicule insertion (One-hour assay) and also the number of contacts with hermaphrodites as well as spicule insertions over a nine hour period (Nine-hours assay). In both assays, we used the same general conditions as in the male mating efficiency assay (particularly as to usage of 14 hermaphrodites), in order to permit comparison of results.

One hour assay
L4 males and L4 hermaphrodites were transferred to separate plates one day prior to testing in order to obtain virgin adult animals. At the beginning of the experiment, one male was placed together with 14 hermaphrodites on a mating plate. We then measured the time until the male touched a hermaphrodite and showed mating behavior (first contact) and until the first time the spicule was inserted (spicule insertion). Observations were terminated after spicule insertion or, if these did not occur, after one hour.

Nine-hours assay
14 L4 hermaphrodites per plate were placed on mating plates one day prior to the experiment. At the same time L4 males were collected and placed on a plate without hermaphrodites. To start the experiment single males were transferred onto the mating plates with the 14 hermaphrodites. Within 9 hours, the plates were inspected 14 times (after 10 minutes, 1 h, 2 h, 4 h, 4.5 h, 5 h, 5.5 h, 6 h, 6.5 h, 7 h, 7.5 h, 8 h, 8.5 h and 9 h). At every inspection, the male was scored as either having no contact with a hermaphrodite, being in contact with a hermaphrodite, or having its spicule inserted.

Self-brood-size assay
This assay served to determine the number of offspring produced trough self-fertilization in the absence of males for two natural isolates, N2 and CB4856. Young adult hermaphrodites, which had no developing embryos in the uterus yet, were placed individually onto plates and moved to new plates at least once every day until they stopped laying eggs. Three days after the removal of the mother the progeny on every plate was counted and summed up to total numbers.

Hermaphrodite outcrossing efficiency assay
This assay was used to test how more frequent mating affects the hermaphrodite's production of self- as well as cross-progeny. Mating plates were prepared four days prior to the assay. One L4 hermaphrodite was placed together with 1, 3, 6 or 12 males onto a mating plate. The hermaphrodite was transferred onto a new mating plate with new young males every day until it stopped laying eggs. The old males were removed from the plates. After one day, additional food bacteria (OP50) were added to the plates and hermaphrodites and males were counted two days later, when they were young adults.

Statistical analysis
All statistical analyses were done with the program JMP IN version 5.1.2 (SAS Insitute Inc., USA) or SPSS version 14.0 (SPSS Inc., USA). The male proportion over time in populations of the different isolates was evaluated using logistic regression based on a full factorial model with time and either population size or $C.$ elegans strain as fixed predictors. The male proportion at specific time points was additionally examined with a Wilcoxon sign rank test. Variation in the number of mated hermaphrodites, the number of offspring per male, and the self- and cross-progeny per mated hermaphrodite was assessed with a General linear model, using an incomplete block design, including male strain, hermaphrodite strain and the interaction between them as fixed factors and experimental block as a random factor. In case of significant interaction terms, significant differences among groups were evaluated with Tukey HSD posthoc tests. Variation in offspring number per repeatedly mated hermaphrodite was tested with an ANOVA, using a full factorial design with male strain and hermaphrodite strain as fixed factors. Subsequent posthoc tests were performed with Tukey HSD. A similar ANOVA was performed to assess the effect of different numbers of males on offspring numbers of repeatedly mated hermaphrodites. In this case, number of males was used as a fixed predictor in the model. With respect to mating behaviour, the time measurements until first contact or first copula were always terminated after 60 min, resulting in non-continuous data. Therefore, differences between crosses were assessed using the Wilcoxon sign rank test. The variation in the number of copulas during this time frame were examined with the Fisher exact test. Differences in the number of contacts or copulations over 14 observation points within a 9 h interval were assessed with the Wilcoxon sign rank test, since the data were non-parametric.

Results and Discussion
Variation in male maintenance among C. elegans strains
We first tested in how far there is variation in male maintenance among populations of different natural C. elegans isolates. Eight strains were tested at two different arbitrarily chosen population sizes (75 and 150). The proportion of males was significantly affected by the factor time, the strain studied, and also the interaction of the two. In par-

ticular, males disappeared completely from the cultures of some strains, among them N2. In contrast, in other strains, among them CB4856, the cultures appeared to reach a stable frequency of males after about two weeks (Fig. 1 and [see Additional files 1, 2]).

To refine the analysis we examined the persistence of males at four different arbitrarily chosen population sizes (40, 70, 100 and 150) for N2 and CB4856 (Fig. 1C and [see Additional files 3, 4]). The proportion of males was significantly reduced within few generations. In fact, males disappeared almost entirely from all N2 cultures. The loss was much slower in CB4856 populations. Here, male frequencies were significantly affected by population size, whereby larger populations (100 and 150) sustained a higher number of males [see Additional file 3]. For these two population sizes, the final male frequency was significantly different between N2 and CB4856, while this was not the case for the smaller populations [see Additional file 4].

Taken together, our results confirm long-standing anecdotal knowledge available within the *C. elegans* community: We, and many others, have noticed that it is necessary to deliberately set up crosses with an excess of males every few generations in order to maintain N2 populations with males for genetic analysis. For CB4856 this is not necessary. Our data are also in agreement with previous studies, in which males were rapidly lost in experimental populations of N2 [7,8,10,11] but maintained at constant levels in those of CB4856 [11]. Since during our experiments the different strains were kept in parallel under identical conditions, the difference in male maintenance between them must have a genetic basis. Thus, our results suggest that *C. elegans* bears considerable intra-specific genetic variation that affect male frequency, making it a potentially selectable trait. Interestingly, population size differences had a significant effect on male persistence. This finding may be a consequence of the population size itself, e.g. smaller populations may loose males more often due to chance, thus accelerating male decline. A non-exclusive alternative explanation may be density differences among the population sizes. In this case, higher densities in the large population sizes may associate with more male-hermaphrodite contacts, which could result in higher mating rates, thus stabilizing male frequencies. At the moment, our results do not allow to distinguish between these two effects.

We decided to further characterize the proximate processes that account for variation in male persistence

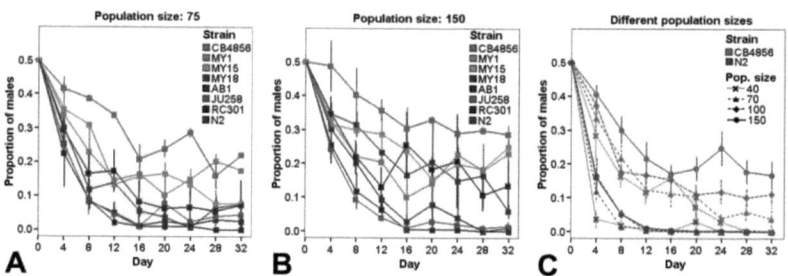

Figure 1
Persistence of males over time in different *C. elegans* strains and population sizes. The proportion of males after the indicated number of days is given. Every four days the populations were reduced to the number specified. Error bars are standard errors. All experiments were started with populations containing approximately 50% males. The first actual measurement was done after the first generation at day 4. A) Decrease of the male frequencies in different wild isolates at population size 75. Each point is the average of two independent measurements. B) Decrease of the male frequencies in different wild isolates at population size 150. Each point is the average of two independent measurements. C) Decrease of the male frequencies in four different population sizes in N2 and CB 4856. Each point is the average of five independent measurements. For more details on results and statistical analysis [see additional files 1, 2, 3, 4].

between two of the extremes, namely the strain N2 and CB4856.

Proximate determinants of male persistence: Male mating efficiency

The factors underlying variation in male maintenance were systematically assessed by reciprocal crosses. The analysis was based on all possible mating combinations between CB4856 and N2, using an incomplete block design (i.e. not all combinations were assayed at the same time; see methods). The main experiment focused on the consequences of repeated mating for individual males, which is likely to be realistic in most populations, where male frequency is usually less than 0.01 [24]. Three parameters were simultaneously evaluated: i) The number of successful copulations per individual male, which was offered an excess of virgin hermaphrodites (14 hermaphrodites) every day over a period of six days, ii) the number of sired offspring per male with – accordingly – virtually unlimited access to mates, and iii) the number of cross- and self-progeny per successfully mated hermaphrodite (for experimental details see Methods). Since early reproduction should have a stronger influence on population dynamics than late reproduction, we also performed a separate analysis of the data from only the first two days (Table 1B). The results lead to essentially identical conclusions like the results for the full reproductive period (Figure 2 and Table 1). Therefore, in the following we focus our discussion on the data for the full reproductive lifespan.

CB4856 males had significantly more successful copulations (Figure 2A and Table 1) and significantly more offspring than N2 males (Figure 2B and Table 1). These two traits were not significantly affected by the hermaphrodite strain used. Therefore, the difference in progeny production by males is most likely a consequence of the differences in mating rates. It does not seem to be caused by variations in the fertilization rates males achieve per successful mating event: The number of cross-progeny per mated hermaphrodite was not significantly affected by any factor of the model or the overall model as a whole ($P > 0.05$; Figure 2C and Table 1). At the same time, it is interesting to note that N2 males appear to produce more cross-progeny with N2 rather than CB4856 hermaphrodites (Fig. 2C; see also similar results obtained after repeated mating of hermaphrodites in Fig. 4A). This effect may account for the trend of a difference produced by the factor male strain in this context (Table 1). It is responsible for the significant interaction term, which was inferred for the data from the first two days (Table 2). One possible explanation for this result is a certain degree of genetic incompatibility, which only becomes visible in one type of cross between the two strains (male N2 and hermaphrodite CB4856) and which may be related to the recent report of genetic incompatibilities among different natural *C. elegans* isolates [33].

Interestingly, N2 hermaphrocites always had significantly more self-progeny than CB4856 – irrespective of the type of the male and irrespective of the cross-progeny produced (Figure 2C and Table 1). This can be explained by their generally higher fertility. In agreement with other authors [15,34] we found in a separate experiment (self-brood-size assay; see methods) that unmated N2 hermaphrodites produced more self-progeny than unmated CB4856 hermaphrodites (284.3 ± 5.7 and 245.3 ± 6.1, respectively; t-test, t_{37} = -4.65, $P < 0.001$). In general, our results on total cross-progeny sired by N2 males (Table 1) as well as total self-progeny by N2 hermaphrodites (see above) are within the range of previously published data [15,34].

Further dissection of mating behavior

In a separate set of experiments we measured different aspects of mating behavior within either the first one or nine hours. As in the above experiments, we combined an individual male with 14 hermaphrodites, although in this case both were always of the same strain (i. e. 1 N2 male × 14 N2 hermaphrodites and 1 CB4856 male × 14 CB4856 hermaphrodites; for further details see Materials and Methods). Within the first one hour, the different crosses did not show any significant variation, neither regarding the time until first male-hermaphrodite contact (253.3 ± 51.4 sec for N2 and 171.5 ± 28.4 sec for CB4856; Wilcoxon test, $Z = 0.95$, $N = 20$ per strain, $P = 0.343$), nor in the time until first successful spicule insertions (2148.5 ± 280.6 sec for N2 and 2212 ± 299.7 sec for CB4856; Wilcoxon test, $Z = 0.10$, $N = 20$ per strain, $P = 0.923$), nor in the number of replicates with successful spicule insertions (12 out of 20 for N2 and 13 out of 20 for CB4856; Fisher exact test, $P > 0.999$). We conclude that the CB4856 males are neither generally better in finding their first mates nor in achieving their first spicule insertion. This result is consistent with the previous finding that N2 and CB4856 show similar behavioral responses to a hermaphrodite-derived cue [12-14].

However, over 14 observation points within the first 9 hours, the CB4856 × CB4856 crosses produced significantly more male-hermaphrodite contacts (Fig. 3 and [see Additional file 5]; Wilcoxon test, $Z = 6.59$, $N = 45$ for CB4856, $N = 47$ for N2, $P < 0.001$) and spicule insertions (Fig. 3 and [see Additional file 5]; Wilcoxon test, $Z = 2.98$, $N = 45$ for CB4856, $N = 47$ for N2, $P = 0.003$). From these observations, we conclude that over time CB4856 males achieve a higher rate of mate contacts and spicule insertions than N2 males. These results are in excellent agreement with the results presented in Fig. 2 and Table 1. Consequently, an overall higher mating frequency could

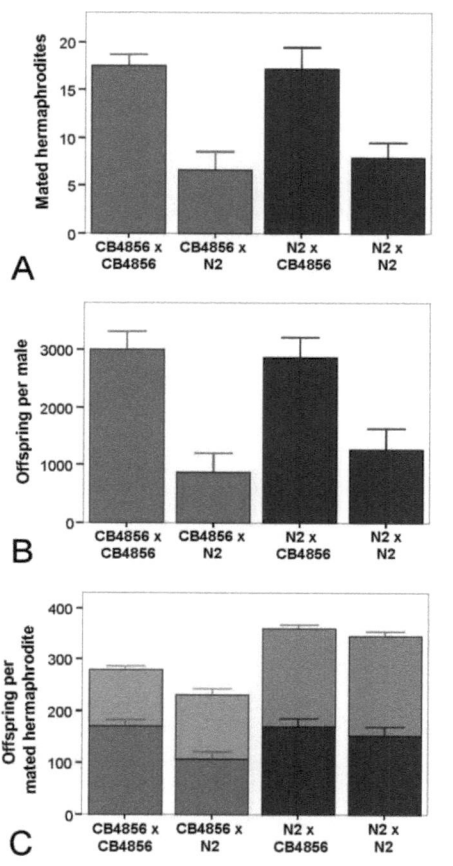

Figure 2
Mating ability and offspring production in reciprocal crosses between N2 and CB4856. A single male was crossed with 14 hermaphrodites and was transferred onto a new plate with 14 young hermaphrodites every day for six days. For crosses the hermaphrodite strain is mentioned first and the male strain second. A) Number of successful copulations a male achieved during its life time. B) Total number of cross-progeny produced per male (estimated as twice the number of males). C) Number of cross- (dark color) and self- (light color) progeny produced per successfully mated hermaphrodite after they were separated from the male. Values are the average of five independent replicates. The error bars designate standard errors. Note: C does not include the progeny these hermaphrodites produced during the time they were on the mating plates. For further details see Materials and Methods.

Table 1: Variation in the number of mated hermaphrodites, offspring per male as well as cross- and self-progeny per mated hermaphrodite for the whole experimental period[a]

Cross	Mates	Offspring/male	Cross-progeny/herm	Self-progeny/herm.
	Mean ± SE	Mean ± SE	Mean ± SE	Mean ± SE
N2 × N2	7.9 ± 1.6	1272.8 ± 367.7	151.2 ± 16.6	192.2 ± 8.8
N2 × CB4856	17.2 ± 2.2	2860.6 ± 341.5	170.6 ± 13.8	190.7 ± 7.2
CB4856 × N2	6.7 ± 1.9	870.9 ± 335.8	106.4 ± 13.8	124.1 ± 11.3
CB4856 × CB4856	17.5 ± 1.2	3003.2 ± 315.1	170.1 ± 11.4	108.1 ± 7.3
Analysis				
Whole model	$F_{6,32} = 5.81$; $P < 0.001$	$F_{6,32} = 4.93$; $P = 0.001$	$F_{6,32} = 2.31$; $P = 0.058$	$F_{6,32} = 16.2$; $P < 0.001$
Male strain	$F_1 = 17.95$, $P < 0.001$	$F_1 = 13.67$, $P < 0.001$	$F_1 = 3.71$, $P = 0.063$	$F_1 = 3.41$, $P = 0.074$
Hermaphrodite strain	$F_1 = 0.23$, $P = 0.636$	$F_1 = 0.02$, $P = 0.888$	$F_1 = 0.56$, $P = 0.460$	$F_1 = 50.53$, $P < 0.001$
Interaction	$F_1 = 0.13$, $P = 0.722$	$F_1 = 0.54$, $P = 0.466$	$F_1 = 2.42$, $P = 0.130$	$F_1 = 0.78$, $P = 0.383$

[a], For each cross (top half of the table), the hermaphrodite strain is given first, the male strain last. The mean number of mated hermaphrodites per male, the mean number of offspring per male, the mean number of cross-progeny per male and mated hermaphrodite as well as the mean number of self-progeny per male and mated hermaphrodite are shown. SE, standard error. Statistical results (bottom half of the table) are shown for the whole model. If the latter shows at least a trend ($P < 0.01$), then the statistical importance of different factors in the model are given. The model also included the random factor "experimental date", which however never produced a significant effect ($F_3 < 2.7$, $P > 0.06$). Significant probabilities are given in bold.

contribute to the observed higher male persistence in CB4856 relative to N2. We cannot explain why N2 males mate less frequently than CB4856. One possibility would be that N2 males require a longer time to replenish their sperm stocks. If so, they could produce a smaller number of sperm over their life time which would explain the reduced number of progeny sired

Variation in male sperm competitiveness (size) could have also affected the observed differences in male persist-

ence. We have not as yet measured this trait. However it is unlikely to have significantly influenced the results on mating efficiency in Fig. 2. We followed the mated hermaphrodites until they ceased to produce progeny, presumably because they had used up their supply of sperm (including both male and hermaphrodite sperm). In this case, all male sperm transferred during mating should have contributed to offspring production irrespective of their competitiveness.

Table 2: Variation in the number of mated hermaphrodites, offspring per male as well as cross- and self-progeny per mated hermaphrodite for the first two days only[a]

Cross	Mates	Offspring/male	Cross-progeny/herm	Self-progeny/herm.
	Mean ± SE	Mean ± SE	Mean ± SE	Mean ± SE
N2 × N2	5.6 ± 1.0	713.0 ± 134.9	135.3 ± 13.2[A,B]	129.4 ± 10.0
N2 × CB4856	11.1 ± 1.3	1283.4 ± 121.0	123.2 ± 12.5[A,B]	157.5 ± 9.7
CB4856 × N2	4.2 ± 0.8	412.4 ± 120.8	84.8 ± 12.0[A]	101.2 ± 13.3
CB4856 × CB4856	9.0 ± 0.5	1192.8 ± 105.8	131.8 ± 9.1[B]	92.4 ± 9.0
Analysis				
Whole model	$F_{6,32} = 5.95$; $P < 0.001$	$F_{6,32} = 5.48$; $P < 0.001$	$F_{6,32} = 2.24$; $P = 0.064$	$F_{6,32} = 6.22$; $P < 0.001$
Male strain	$F_1 = 13.88$, $P < 0.001$	$F_1 = 12.68$, $P = 0.001$	$F_1 = 0.86$, $P = 0.360$	$F_1 = 0.28$, $P = 0.600$
Hermaphrodite strain	$F_1 = 0.90$, $P = 0.351$	$F_1 = 0.47$, $P = 0.499$	$F_1 = 0.55$, $P = 0.464$	$F_1 = 10.02$, $P = 0.003$
Interaction	$F_1 = 0.19$, $P = 0.668$	$F_1 = 0.66$, $P = 0.421$	$F_1 = 6.44$, $P = 0.016$	$F_1 = 3.73$, $P = 0.063$

[a], For each cross (top half of the table), the hermaphrodite strain is given first, the male strain last. The mean number of mated hermaphrodites per male, the mean number of offspring per male, the mean number of cross-progeny per male and mated hermaphrodite as well as the mean number of self-progeny per male and mated hermaphrodite are given. SE, standard error. Statistical results (bottom half of the table) are shown for the whole model. If the latter shows at least a trend ($P < 0.01$), then the statistical importance of different factors in the model are given. In case of a significant interaction factor (cross-progeny per mated hermaphrodite), we also provide the results of Tukey HSD posthoc tests, whereby significantly different groups are indicated by different superscript Captial letters in the top part of the table. The model also included the random factor "experimental date", which did not produced a significant effect ($F_3 < 1$, $P > 0.4$) except of the analysis of self-progeny per mated hermaphrodites ($F_3 = 3.03$, $P = 0.044$). Significant probabilities are given in bold.

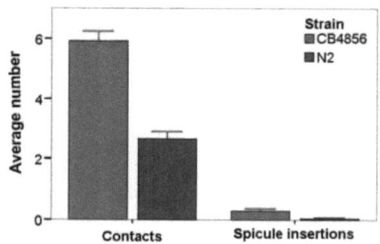

Figure 3
Number of contacts and spicule insertions observed over 14 observation points within 9 hours. Mating assays with one male and 14 hermaphrodites were set up for CB4856 and N2, using 45 and 47 replicates, respectively. The plates were inspected 14 times within the first 9 hours. The figure shows the average number of male-hermaphrodite contacts and spicule insertions. Each spicule insertion was also considered to be a contact. The error bars designate standard errors. For exact numbers and statistical analysis [see additional file 5].

Consequence of repeated mating of hermaphrodites
In the experiments leading to Figure 2 and Table 1 it is likely that individual hermaphrodites mated only once or very few times since they were in excess and they were removed from the males after one day. Given the low number of males in natural *C. elegans* populations, mating only once, if at all, might be realistic. However, in our male maintenance assays (see Figure 1), the male frequency was initially set to approximately 0.5, thus allowing for repeated mating interactions. Therefore, we asked whether repeated mating could influence the number of self- as well as cross-progeny. For this purpose, we set up all reciprocal crosses between N2 and CB4856 and, for each cross, single hermaphrodites were mated each day with a virgin male until the hermaphrodite ceased to produce progeny (see Materials and Methods, hermaphrodite outcrossing efficiency assay with one male).

In agreement with earlier literature [34], we found that repeated mating (repeated mating assay with one male, see Methods) increased the total number of progeny and the proportion of cross-progeny (Figure 4A and [see Additional file 6]). N2 hermaphrodites produced significantly more progeny than CB4856 hermaphrodites (ANOVA, hermaphrodite strain effect, $F_1 = 107.69$, $P < 0.001$) and the proportion of self-progeny was small. The origin of the male did not make any significant difference (ANOVA, male strain effect, $F_1 = 2.93$, $P = 0.108$). Analysis of the first two days only produced essentially identical results [see Additional file 7]. Consequently, in the case of repeated mating of hermaphrodites, N2 would produce a significantly larger number of cross-progeny and thus a larger proportion of males than CB4856.

At first sight, this result appears to contradict our findings from the male maintenance assays, where N2 populations loose males rapidly in contrast to CB4856. However, hermaphrodites in the repeated mating experiment were exposed to "new" virgin males every day (most likely with high mating efficiency), whereas hermaphrodites in the male maintenance assay encounter the same males over consecutive days, which are likely to show reduced mating efficiency over time (due to the likely high energy demand per mating as well as sperm depletion). The reduction in mating efficiency is particularly pronounced in N2 (see results from the mating behaviour experiments above). Therefore, we expect fewer matings, a relatively larger number of self-progeny, and thus a continuous decrease of males in the N2 populations over time.

A possible higher cost of multiple mating may further contribute to more rapid initial male decline in N2 in the male maintenance assays. Repeated mating with increasing numbers of males caused a significant reduction in offspring number in both strains (Figure 4B and [see Additional file 8]). Importantly, this reduction was more pronounced for N2, for which it caused a loss of up to 53% progeny compared to a maximum of about 40% for CB4856 (comparison between repeated mating with 1 versus 12 males). Consequently, if repeated mating of hermaphrodites did occur in the experimental populations, then the higher costs (i.e. offspring loss) of copulations with multiple males for N2 should decrease male frequencies and thus outcrossing rates to a larger extent in N2 than CB4856.

The finding of a cost of multiple mating is likely a consequence of sexual conflict, as reported for a large diversity of organisms [35,36]. One possible explanation for this observation could be increased intra-sexual male-male competition or otherwise detrimental male-male interactions, which were shown in the past to decrease *C. elegans* male life-span [37]. Alternatively, it could result from inter-sexual antagonisms such as those mediated by male manipulative substances that are transferred during copulation, in order to enhance male fertilization success [35,36]. In *C. elegans*, the possible relevance of such inter-sexual conflict was previously indicated by reduced hermaphrodite longevity after mating [38].

Conclusion
Our experiments suggest that the combination of two traits are likely involved in determining the difference in

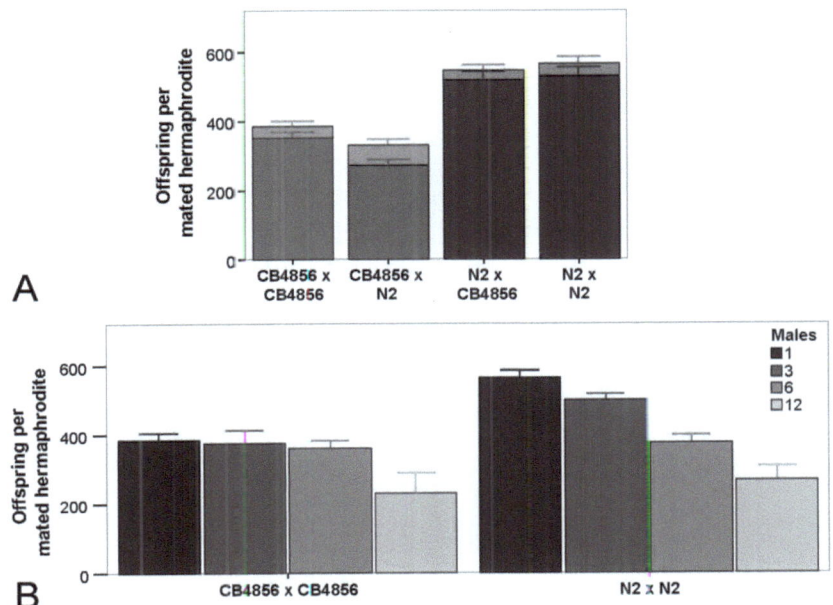

Figure 4
Offspring production by repeatedly mated hermaphrodites. A) Repeated mating of an individual hermaphrodite with a single virgin male every day. The hermaphrodite strain is indicated first and the male strain second. The number of cross-progeny (dark colors) and self-progeny (light colors) were determined. Each value is the average of five replicates. The error bars represent standard errors. For exact numbers and statistical analysis [see additional file 7]. Repeated mating of individual hermaphrodites with different numbers of virgin males added every day. The figure shows total progeny. Note that the proportion of self-progeny was low (see A) and did not differ significantly between treatments; thus, the observed variation is mainly determined by cross-progeny. Each value is the average of five hermaphrodites. The error bars represent standard errors. For exact numbers and statistical analysis [see additional file 8].

male maintenance between CB4856 and N2: i) CB4856 males achieved a larger number of successful copulations and therefore sired more cross-progeny than N2. Consequently, a mixed-gender CB4856 population will contain a larger number of cross-fertilized hermaphrodites and thus produce more males than a corresponding N2 population. The resulting higher frequency of males should further enhance cross-fertilization rates in CB4856, because male density positively links with hermaphrodite mating rates [9]. ii) Unmated and singly mated N2 hermaphrodites produced a higher number of self-progeny than corresponding CB4856 hermaphrodites. This parameter reduces male density and thus mating rates in N2, which additionally amplifies the loss of males within N2 populations.

Authors' contributions
VW did all the bench work and participated in the experimental design, the analysis of the data and the writing of the manuscript. HS participated in the experimental design and the data analysis. He did all the statistical analyses and he co-wrote the manuscript together with AS. AS participated in the experimental design and the data analysis. He coordinated the whole work and supervised the

practical work. He co-wrote the manuscript together with HS. The contributions of HS and AS should be considered equal. All Authors read and approved the final manuscript.

Additional material

Additional file 1
Supplementary table 1. Logistic regression of male proportion in different natural isolates.
Click here for file
[http://www.biomedcentral.com/content/supplementary/1472-6785-8-12-S1.doc]

Additional file 2
Supplementary table 2. Male proportion averaged over days 16 to 32 for different strains and two population sizes.
Click here for file
[http://www.biomedcentral.com/content/supplementary/1472-6785-8-12-S2.doc]

Additional file 3
Supplementary table 3. Logistic regression of male proportion in N2 and CB4856 with different population sizes.
Click here for file
[http://www.biomedcentral.com/content/supplementary/1472-6785-8-12-S3.doc]

Additional file 4
Supplementary table. Male proportion on day 32 for different population sizes of the strains N2 and CB4856.
Click here for file
[http://www.biomedcentral.com/content/supplementary/1472-6785-8-12-S4.doc]

Additional file 5
Supplementary table 5. Variation in the number of contacts and spicule insertions within the first 9 hours.
Click here for file
[http://www.biomedcentral.com/content/supplementary/1472-6785-8-12-S5.doc]

Additional file 6
Supplementary table 6. Variation in the number of cross- and self-progeny per repeatedly mated hermaphrodite for the whole experimental period.
Click here for file
[http://www.biomedcentral.com/content/supplementary/1472-6785-8-12-S6.doc]

Additional file 7
Supplementary table 7. Variation in the number of cross- and self-progeny per repeatedly mated hermaphrodite for the first two days only.
Click here for file
[http://www.biomedcentral.com/content/supplementary/1472-6785-8-12-S7.doc]

Additional file 8
Supplementary table 8. Variation in total hermaphrodite offspring number after repeated mating to either 1, 3, 6, or 12 males.
Click here for file
[http://www.biomedcentral.com/content/supplementary/1472-6785-8-12-S8.doc]

Acknowledgements
Some of the *C. elegans* strains were supplied by the *Caenorhabditis* Genetics Center, which is funded by the National Institutes of Health National Center for Research Resources. We thank Drs Robbie Rae and Matthias Herrmann and Nadine Timmermeyer for critical reading of the manuscript. This work was funded by the Max Planck Society and grant SCHU1415/5-1 from the Deutsche Forschungsgesellschaft to HS.

References
1. Hope IA: **Background on *Caenorhabditis elegans*.** In *C. elegans a practical approach* Edited by: Hope IA. Oxford: Oxford University Press; 1999:1-15. [Hames BD (Series Editor): The Practical Approach series]
2. Wood WB: **Introduction to *C. elegans* Biology.** In *The Nematode Caenorhabditis elegans* Edited by: Wood WB. New York: Cold Spring Harbor Laboratory Press; 1988:1-16.
3. LaMunyon CW, Ward S: **Larger sperm outcompete smaller sperm in the nematode *Caenorhabditis elegans*.** *Proc Biol Sci* 1998, **265:**1997-2002.
4. LaMunyon CW, Ward S: **Evolution of sperm size in nematodes: sperm competition favours larger sperm.** *Proc Biol Sci* 1999, **266:**263-267.
5. Bell G: *Masterpiece of Nature: The evolution of sexuality* Berkeley: University of California Press; 1982.
6. Maynard-Smith J: *The evolution of sex* Cambridge, UK: Cambridge University Press; 1978.
7. Chasnov JR, Chow KL: **Why are there males in the hermaphroditic species *Caenorhabditis elegans*?** *Genetics* 2002, **160:**983-994.
8. Stewart AD, Phillips PC: **Selection and maintenance of androdioecy in *Caenorhabditis elegans*.** *Genetics* 2002, **160:**975-982.
9. Cutter AD, Aviles L, Ward S: **The proximate determinants of sex ratio in *C. elegans* populations.** *Genet Res* 2003, **81:**91-102.
10. Cutter AD: **Mutation and the experimental evolution of outcrossing in *Caenorhabditis elegans*.** *J Evol Biol* 2005, **18:**27-34.
11. Teotonio H, Manoel D, Phillips PC: **Genetic variation for outcrossing among *Caenorhabditis elegans* isolates.** *Evolution* 2006, **60(6):**1300-1305.
12. Chasnov JR, So WK, Chan CM, Chow KL: **The species, sex, and stage specificity of a *Caenorhabditis* sex pheromone.** *Proc Natl Acad Sci USA* 2007, **104:**6730-6735.
13. Simon JM, Sternberg PW: **Evidence of a mate-finding cue in the hermaphrodite nematode *Caenorhabditis elegans*.** *Proc Natl Acad Sci USA* 2002, **99:**1598-1603.
14. Garcia LR, LeBoeuf B, Koo P: **Diversity in mating behavior of hermaphroditic and male-female *Caenorhabditis* nematodes.** *Genetics* 2007, **175:**1761-1771.
15. Hodgkin J, Doniach T: **Natural variation and copulatory plug formation in *Caenorhabditis elegans*.** *Genetics* 1997, **146:**149-164.
16. Manoel D, Carvalho S, Phillips PC, Teotonio H: **Selection against males in *Caenorhabditis elegans* under two mutational treatments.** *Proc Biol Sci* 2007, **274:**417-424.
17. Agrawal AF, Lively CM: **Parasites and the evolution of self-fertilization.** *Evolution* 2001, **55(5):**869-879.
18. Fischer RA: *The genetical theory of natural selection* Oxford: Clarendon Press; 1930.
19. Hamilton WD, Axelrod R, Tanese R: **Sexual reproduction as an adaptation to resist parasites (a review).** *Proc Natl Acad Sci USA* 1990, **87:**3566-3573.
20. Kondrashov AS: **Deleterious mutations and the evolution of sexual reproduction.** *Nature* 1988, **336:**435-440.

21. Muller HJ: **Some genetic aspects of sex.** *American Naturalist* 1932, **66:**118-138.
22. Muller HJ: **The relation of recombination to mutational advantage.** *Mutation Research* 1964, **1:**2-9.
23. Barrière A, Félix MA: **High local genetic diversity and low outcrossing rate in *Caenorhabditis elegans* natural populations.** *Curr Biol* 2005, **15:**1176-1184.
24. Barrière A, Félix MA: **Temporal dynamics and linkage disequilibrium in natural *Caenorhabditis elegans* populations.** *Genetics* 2007, **176:**999-1011.
25. Cutter AD: **Nucleotide polymorphism and linkage disequilibrium in wild populations of the partial selfer *Caenorhabditis elegans*.** *Genetics* 2006, **172:**171-184.
26. Haber M, Schungel M, Putz A, Müller S, Hasert B, Schulenburg H: **Evolutionary history of *Caenorhabditis elegans* inferred from microsatellites: evidence for spatial and temporal genetic differentiation and the occurrence of outbreeding.** *Mol Biol Evol* 2005, **22:**160-173.
27. Sivasundar A, Hey J: **Population genetics of *Caenorhabditis elegans*: the paradox of low polymorphism in a widespread species.** *Genetics* 2003, **163:**147-157.
28. Sivasundar A, Hey J: **Sampling from natural populations with RNAi reveals high outcrossing and population structure in *Caenorhabditis elegans*.** *Curr Biol* 2005, **15:**1598-1602.
29. Charlesworth B, Charlesworth D: **Some evolutionary consequences of deleterious mutations.** *Genetica* 1998, **102–103:**3-19.
30. Pannell JR: **The evolution and maintenance of Androdioecy.** *Annu Rev Ecol Syst* 2002, **33:**397-425.
31. Stiernagle T: **Maintenance of *C. elegans*.** In *C elegans a practical approach* Edited by: Hope IA. Oxford: Oxford University Press; 1999:51-67. [Hames BD (Series Editor): *The Practical Approach series*]
32. [http://www.cbs.umn.edu/CGC/].
33. Dolgin ES, Charlesworth B, Baird SE, Cutter AD: **Inbreeding and outbreeding depression in *Caenorhabditis* nematodes.** *Evolution* 2007, **61(6):**1339-1352.
34. Hodgkin J, Barnes TM: **More is not better: brood size and population growth in a self-fertilizing nematode.** *Proc Biol Sci* 1991, **246:**19-24.
35. Arnquist G, Lowe L: *Sexual conflict* Princeton and Oxford: Princeton University Press; 2005.
36. Chapman T, Arnquist G, Bangham J, Rowe L: **Sexual conflict.** *Trends in Ecology and Evolution* 2003, **18:**41-47.
37. Gems D, Riddle DL: **Genetic, behavioral and environmental determinants of male longevity in *Caenorhabditis elegans*.** *Genetics* 2000, **154:**1597-1610.
38. Gems D, Riddle DL: **Longevity in *Caenorhabditis elegans* reduced by mating but not gamete production.** *Nature* 1996, **379:**723-725.

Do males facilitate the spread of novel phenotypes within populations of the androdioecious nematode *Caenorhabditis elegans*?

VIKTORIA WEGEWITZ,[1] HINRICH SCHULENBURG,[2] ADRIAN STREIT[1]

Abstract: In the androdioecious nematode *Caenorhabditis elegans*, self-fertilization is the predominant mode of reproduction. Nevertheless, males do occur, and it is still unclear if these represent a selective advantage or merely an evolutionary relict. In this study, we first tested the hypothesis that the production of males might benefit invaders to resident populations. We added single, GFP-marked worms to established laboratory populations and followed GFP frequencies over time. Mated hermaphrodites and also males were more successful in invading resident populations if compared to single, unmated hermaphrodites. The observed higher frequencies should increase the likelihood that any of the associated invading alleles persist. Second, we tested the hypothesis that males and, thus, higher outcrossing rates, are specifically favored under changing environmental conditions. After an outbred population was subjected to changing stress or to control laboratory conditions, we measured the male maintenance of the resulting populations. Interestingly all populations, experimental and control alike, showed high male maintenance, suggesting that persistence of males is also favored under standard laboratory conditions.

Key words: *Caenorhabditis elegans*, ecology, hermaphrodite, male, mode of reproduction.

Androdioecious reproductive systems consisting primarily of hermaphrodites, instead of females and males, exist in multiple phyla (for an overview see Stewart and Phillips, 2002). From an evolutionary perspective, androdioecy is puzzling, and the replacement of females with hermaphrodites appears only possible under certain rather strict conditions (Charnov et al., 1976). Further, once hermaphrodites are present and can reproduce by self-fertilization, males appear superfluous. In analogy with the two-fold cost of males in theories on the evolution of sex (Maynard-Smith, 1978; Bell, 1982), males might even represent a burden, decreasing individual fitness.

The arguably best-characterized androdioecious species is *Caenorhabditis elegans*. *C. elegans* hermaphrodites reproduce either by self-fertilization or by cross-breeding with a male [for a general introduction see Wood (1988) and Hope (1999)]. Cross-fertilization between hermaphrodites does not occur. Hermaphrodites are essentially females that initially produce a limited quantity of sperm and then switch to the production of eggs for the rest of their reproductive life. The sperm is stored in the spermatheca and can be used to fertilize the eggs after the sex switch of the germ line. Sex is determined genetically by the presence of two (hermaphrodites) or one (males) X chromosomes along with five pairs of autosomes. Consequentially, almost the entire self-progeny is hermaphroditic. The very few males that arise spontaneously in the self-progeny of hermaphrodites (around 0.2% in the standard laboratory strain N2) are the result of X chromosome non-disjunction events. Half of the sperm produced by males contain no X chromosome, and as a consequence, 50% of the progeny sired by males is male. Upon mating, male-derived sperm is also stored in the spermatheca along with the hermaphrodite's own sperm. Usually, male sperm has a competitive advantage over the hermaphrodite's sperm, at least in part due to its larger size (LaMunyon and Ward, 1998, 1999, 2002).

It has been suggested that *C. elegans* males might merely be evolutionary relicts with no particular function, which are still present after a relatively recent evolution of females into self-fertilizing hermaphrodites (Chasnov and Chow, 2002). Indeed, hermaphroditism in all three contemporary hermaphroditic species within the genus *Caenorhabditis* has likely arisen independently, and all of these species have close gonochoristic relatives, indicating that the transition to self-fertilization happened relatively recently (Kiontke et al., 2004; K. Kiontke and D. H. Fitch, NYU, pers. com.). A similar picture emerges in *Pristionchus*, another well-studied nematode genus, where hermaphroditism arose at least six times independently (Mayer et al., 2007; W. E. Mayer, M. Herrmann and R. J. Sommer, MPI Dev. Biol., pers. com.). However, given that "relatively recently" in this context still means up to tens of millions of years (Kiontke et al., 2004; Mayer et al., 2007), it is striking that males still exist in all known hermaphroditic species of *Caenorhabditis* (K. Kiontke and D. H. Fitch, NYU, pers. com.) and *Pristionchus* (M. Herrmann, W. E. Mayer and R. J. Sommer, MPI Dev. Biol., pers. com.). Indeed, androdioecy may be maintained by as yet unknown selective forces, which prevent the complete loss of males as well as a switch back to a dioecous reproduction system (Stewart and Phillips, 2002). It has been estimated that a purely selfing *C. elegans* population would be driven to extinction within less than a million year by the accumulation of slightly deleterious mutations (Loewe and Cutter, 2008).

A first prerequisite for males to play a role in populations is that they, and consequentially out-crossing, exist at a level that significantly influences population genetics. In cultures of the standard laboratory strain N2 that are initiated with high numbers of males, the

---

Received for publication September 2, 2009.
[1]Department of Evolutionary Biology, Max Planck Institute for Developmental Biology, Spemannstrasse 35, D-72076 Tübingen, Germany.
[2]Zoological Institute, Christian Albrecht University at Kiel, Am Botanischen Garten 1-9, D-24118 Kiel, Germany.
Some of the *C. elegans* strains were supplied by the *Caenorhabditis* Genetics Center, which is funded by the National Institutes of Health National Center for Research Resources. The contributions of HS and AS to this manuscript should be considered equal. This work was funded by the Max Planck Society and grant SCHU1415/5-1 from the Deutsche Forschungsgesellschaft to HS.
E-mail: adrian.streit@tuebingen.mpg.de
This paper was edited by Amy Treonis.

male frequency declines rapidly, and most of the time males disappear from the populations within less than 20 generations (Chasnov and Chow, 2002; Stewart and Phillips, 2002; Cutter et al., 2003; Cutter, 2006; Teotonio et al., 2006; Wegewitz et al., 2008). However, other natural isolates behave differently if assayed under the same standard laboratory conditions and maintain males over longer periods of time (Teotonio et al., 2006; Wegewitz et al., 2008). Recently, several authors have attempted to infer the outcrossing frequencies in natural populations by measuring linkage disequilibrium, heterozygosity, or genetic diversity (Sivasundar and Hey, 2003; Barriere and Felix, 2005; Haber et al., 2005; Sivasundar and Hey, 2005; Cutter, 2006; Barriere and Felix, 2007). All these studies support the notion that outcrossing does occur in wild populations, and that males do leave an appreciable genetic footprint in natural populations. Except for Sivasundar and Hey (2005), who estimated an outcrossing rate of 0.2, it is broadly accepted that outcrossing is rare in natural populations, ranging in between 10^{-5} and 0.02. Nevertheless, even these rare outcrossing events may be sufficient to reduce the mutational load and/or maintain sufficient genetic diversity required for rapid adaptation to fluctuating environments (Charlesworth and Charlesworth, 1998; Agrawal and Lively, 2001; Pannell, 2002). Laboratory evolution experiments suggested that elevated mutation rates induced either by chemical mutagens (Manoel et al., 2007) or by a deficient DNA repair mechanism (Cutter, 2005) represent a selective force for higher male frequencies. Very recently, it has been shown that outcrossing was not only favored under conditions of increased mutation rate but also during the adaptation to the presence of a pathogen (Morran et al., 2009b). These observations are in agreement with the expectation based on theoretical considerations, according to which males and frequent outcrossing are beneficial under variable environmental conditions and/or high deleterious mutation rates (Fischer, 1930; Muller, 1932; Muller, 1964; Kondrashov, 1988; Hamilton et al., 1990; Agrawal and Lively, 2001). In this context, it is interesting to note that *C. elegans* also appears to plastically increase the outcrossing rate in response to stressful conditions (Morran et al., 2009a).

In a previous study, we showed that in a situation with virtually unlimited access to hermaphrodites a male can produce a considerably higher number of progeny than a hermaphrodite, which reproduces by self-fertilization (Wegewitz et al., 2008). This effect was much more pronounced in the strain CB4856 where males sired more than 10 times as many progeny as unmated hermaphrodites produced. Even the "poorly" mating N2 males still gave rise to more than three times as many progeny as unmated hermaphrodites of the same strain. However, males needed to mate with multiple hermaphrodites to reach this reproductive success. From these numbers, one would expect that a male that arose spontaneously or invaded into a population of hermaphrodites would contribute more to the gene pool of the next generation than any of the hermaphrodites, provided the population density is high enough that the male finds multiple mates during its life. In order to test this prediction, we asked if particular phenotypes may spread and persist more easily if they invade a resident, largely hermaphrodite population as male individuals or mated hermaphrodites rather than virgin (and thus exclusively selfing) hermaphrodites. We performed experiments where we added single individuals that were marked with a transgene to stable populations of unmarked worms and followed the frequency of the transgenic phenotype.

In a second experiment we asked if varying non-mutagenic stress conditions also act as selective pressure in favor of higher male frequencies, because these could possibly enhance the spread of novel advantageous phenotypes. To create a starting population with the genetic potential to achieve various levels of male maintenance, we interbred N2 and CB4856, two strains on the low and the high ends, respectively, of the spectrum of male maintenance found in natural isolates (Wegewitz et al., 2008). The resulting hybrid populations were subjected to varying environmental conditions (high salt, low and high temperatures, pathogenic bacteria, and standard laboratory conditions) or continuous standard laboratory conditions.

MATERIALS AND METHODS

C. elegans cultures: *C. elegans* was cultured on NGM plates with *Escherichia coli* strain OP50 as food (Stiernagle, 1999). Mating plates consisted of 6 cm NGM plates seeded in the center with 30 μl of an *E. coli* (OP50) culture. Cultures were incubated in an air-conditioned room at a temperature of 21±1°C and 40% humidity. Cultures were kept in boxes, randomized in piles that were evenly distributed within the boxes.

Strains used: N2: Standard laboratory wild type strain, isolated in Bristol, UK CB4856: Standard polymorphic mapping strain isolated in Hawaii. PD4792: *mIs11[myo-2::gfp + pes-10::zfp + gut::gfp]* All three strains were requested from the *Caenorhabditis* Genetics Center at the University of Minnesota (http://biosci.umn.edu/CGC/). QA351 *ytIs3[sur-5::gfp]* (created by microinjection and UV induced integration (Jin, 1999) of pTG96 (Gu et al., 1998) followed by five back crosses with N2). QA353: *mIs11[myo-2::gfp + pes-10::gfp + gut::gfp]* IV (created by backcrossing PD4792 to N2 five times) QA354: *mIs11[myo-2::gfp + pes-10::gfp + gut::gfp]* IV (created by backcrossing PD4792 to CB4856 five times).

Male maintenance assays: Male maintenance assays were done with a population size of 150 as described by Wegewitz et al., (2008).

Invasion experiments: "Stable populations" were started with N2 hermaphrodites and maintained by transferring a fixed number (population size) of worms to

a new plate every three days, without paying attention to developmental stage, sex, or GFP fluorescence. To transfer the worms, they were washed off the old plate with M9 solution, and the number of animals in an aliquot were counted. Based on this count, the total number of worms on the plate was estimated, and the appropriate number of worms was pipetted onto a new plate. To initiate the experiment, one individual that was genetically marked with *myo-2::gfp* (invader) was added to each "stable population" immediately after a transfer. The invaders were either virgin hermaphrodites or mated hermaphrodites or males. The populations were maintained as described above for eight transfers. Prior to each transfer and at the end of the experiment, the number of GFP positive worms on each plate was counted. As invaders we used QA351 and QA353 (essentially the genetic background of N2) and QA354 (essentially the genetic background of CB4856).

For Invasion Experiment 1, the population size was 500. Four replicates for each type of invader were done in parallel, and the experiment was repeated twice with QA351 and twice with QA353, resulting in two times eight replicates per type of invader. For Invasion Experiment 2, population sizes of 100 and 500 and with QA353 and QA354 as invaders were used, resulting in 12 different treatments. One replicate for each of the treatments were carried out simultaneously, and the experiment was repeated six times.

Experimental Evolution Experiment: The different environmental conditions were:

1) High Salt: NGM/OP50 plates containing 20 g/l NaCl at 20°C.
2) Low temperature: NGM/OP50 plates at 15°C.
3) High temperature: NGM/OP50 plates at 25°C
4) Pathogen: NGM plates seeded four days prior to use with 1 ml of a 5:2 mixture of CBX102 (*Microbacterium nematophilum*) and *E. coli* OP50 overnight cultures at 20°C.
5) Control: NGM/OP50 plates at 20°C.

Populations were initiated by placing 10 N2 and 10 CB4856 hermaphrodites on mating plates together with 30 males of the other strain for 24 h. Then the hermaphrodites from each cross were transferred to a 10 cm NGM/OP50 plate and allowed to reproduce for three days. The worms were washed from the plates with M9 buffer (Stiernagle, 1999), and from each cross 60 individuals were placed on five high salt or five control plates without paying attention to developmental stage or sex, resulting in five replicates for selection (series B) and five control replicates (series A) with starting populations of 120 individuals. Another five selection (series D) and control (series C) replicates were initiated one day later.

After four days 120 individuals from each plate were transferred to new plates without paying attention to developmental stage or sex and subjected to low temperature (selection) or control conditions.

After four days, 120 individuals from each plate were transferred to new plates and subjected to high temperature (selection) or control conditions.

After four days, 120 individuals from each plate were transferred to plates containing pathogenic bacteria (selection) or control plates. The reminder of the cultures was frozen as described by (Stiernagle, 1999).

After four days, all cultures were treated with hypochloride (Stiernagle, 1999) to remove the pathogenic bacteria, and the isolated embryos were allowed to hatch on plates without food for one day. Then, 120 larvae were placed on NGM/OP50 plates. Two days later, three N2 and three CB4856 males were added to each culture to avoid statistical loss of genotypes. One day later, 120 individuals per culture were placed on NaCl (selection) or control plates to start the second round of selection. In total, five rounds were performed. The experiment was terminated with the freezing step of the fifth round. For the further analysis, aliquots of the cultures were thawed, and the male maintenance test started within two generations.

"Chunking" Experiment: The whole experiment was performed on 10 cm NGM/OP50 plates. The cultures were started by combining a total of 10 hermaphrodites and, if applicable, 10 males on one plate as specified below.

Mixed: hermaphrodites: Five progeny of a cross N2 x CB4856 plus 5 progeny of a cross CB4856 x N2; males: Five progeny of a cross N2 x CB4856 plus 5 progeny of a cross CB4856 x N2.

N2 with males: hermaphrodites: 10 N2; males: 10 N2.

CB4856 with males: hermaphrodites: 10 CB4856; males: 10 CB4856.

N2 no males: hermaphrodites: 10 N2; males: none.

CB4856 no males: hermaphrodites: 10 CB4856; males: none.

For each combination, three replicates were performed. The cultures were incubated at 20°C. After seven days, a square (chunk) of 16 x 16 mm was cut out from the agar about one third of a plate radius away from the center and transferred upside down onto a new plate (chunking). The chunking step was repeated every seven days until a total of 12 transfers were completed. At the end, the worms were frozen. Aliquots of the cultures were thawed for further analysis, and the male maintenance test started within two generations.

Statistical analysis: The data were analysed using generalized linear models, based on logistic regression analysis. The analyses were performed with the program JMP 8.0 (SAS Institute Inc.), and all graphs were produced with Sigmaplot 11.0 (Systat Software Inc.).

For the invasion experiments, the model included the following factors: treatment (i.e. addition of hermaphrodites, males, or mated hermaphrodites), day of measurement, the interaction between day and treatment, and block (i.e. date when a particular combination of experiments was started). The response variable

was the proportion of GFP-positive nematodes within the population. We assessed the overall model and then separately several models, which specifically evaluated the difference between two of the treatment alternatives (e.g., addition of hermaphrodites versus addition of males). In all cases, likelihood ratio effect tests were used to evaluate the impact of a particular factor on the variance of the data. The significance level was adjusted using the false-discovery rate (FDR) to account for multiple testing. For Invasion Experiment 1, we performed separate analyses for two types of crosses (i.e. QA351 vs. QA353). For Invasion Experiment 2, we performed separate analyses for the two population sizes (i.e. 100 vs. 500) and the two types of crosses (i.e. QA354 vs. QA353).

For the Experimental Evolution and the Chunking experiments, the models included the following factors: treatment (i.e. the different types of crosses and strains), day of measurement, interaction between treatment and day, and parental population (from which the nematodes were taken). We again assessed an overall model and then several separate models, which particularly addressed the difference between two treatments. Likelihood ratio effect tests were used to test the impact of a factor and FDR to account for multiple testing.

RESULTS AND DISCUSSION

We added individual worms marked with a *gfp* reporter gene (strains QA351 and QA353) to plates with 500 hermaphrodites to mimic an event of an invasion or the occurrence of a spontaneous mutation (Invasion Experiment 1). Consistent with the expectation, *gfp* positive worms reached higher frequencies when the reporter gene was brought to the population through a male or a mated hermaphrodite than through a non-mated hermaphrodite (Fig. 1, Table 1). Next we repeated the experiment with QA353 with two different population sizes (100 and 500), and we also included QA354, a strain with essentially the genetic background of the more efficiently mating strain CB4856 (Wegewitz et al., 2008) as invader (Fig. 2, Table 2). In all four treatments, mated hermaphrodites did significantly better than non-mated hermaphrodites. Interestingly, males did better in the larger populations of 500 than in the smaller populations of 100 individuals. A possible explanation for this is that the reproductive success of the males might have been limited by a low rate of finding suitable mates (young adult hermaphrodites) in the 100-individuals mixed-stage populations. Males must mate with multiple hermaphrodites to maximize their reproductive potential (Wegewitz et al., 2008), and it has been suggested that mate encounter rates are an important factor for male reproductive success (Lopes et al., 2008). In contrast, hermaphrodites – mated or unmated – could immediately reproduce and thus contribute to the next generation of these small populations.

For the purpose of this experiment, we considered the *myo-2::gfp* a neutral genetic marker. Although we did not observe any obvious deleterious effects of the transgene, we cannot exclude a slight selective disadvantage for worms that carry the marker. However, this effect would have been the same for all worms that were compared directly (i.e. the transgenic males, the transgenic unmated hermaphrodite, and the mated hermaphrodite). Therefore, while the transgene might have slightly affected the absolute numbers, it would not have compromised the comparison. As expected after the addition of a single invader, the frequency of the introduced gene remained rather low, and in several cases it disappeared across time through genetic drift, especially if the invader was a virgin hermaphrodite. In particular, the GFP marker disappeared by the end of the experiment in all

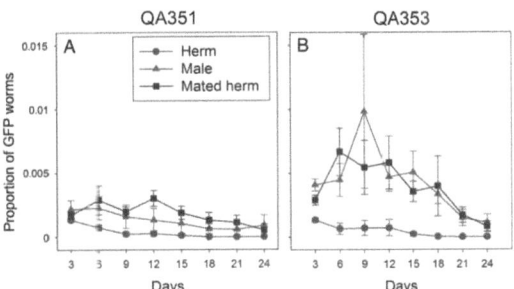

FIG. 1. Invasion Experiment 1: Proportion of GFP-positive worms (Y-axis) over time (X-axis) after the addition of a single GFP marked (QA351 [A] or QA353 [B]) virgin hermaphrodite (circles), male (triangles) or mated hermaphrodite (squares) to a population of 500 N2 worms on day 0. Every three days the populations were reduced to 500. GFP-positive worms were counted immediately before the reduction of the population. The error bars designate standard errors. Each point is the average of 8 independent measurements. For the statistical analysis see Table 1.

TABLE 1. Statistical analysis of the effect of treatment on the proportion of GFP-positive offspring in Invasion Experiment 1.

Cross	Comparison	$\chi^2_{df=1}$	P
QA351	Overall model	79.6	< 0.0001[a]
	Herm. vs. Male	15.2	< 0.0001[a]
	Herm vs. Mated herm.	97.3	< 0.0001[a]
	Male vs. Mated herm.	21.4	< 0.0001[a]
QA353	Overall model	117.0	< 0.0001[a]
	Herm. vs. Male	89.2	< 0.0001[a]
	Herm vs. Mated herm.	93.6	< 0.0001[a]
	Male vs. Mated herm.	0.5	0.4769

All models were significantly better than a minimal model ($\chi^2_{df=4,6} \geq 24.9$, $P < 0.0001$). The impact of the treatment factor was evaluated with a likelihood ratio effect test.

[a] Significant probabilities (P) according to the false-discovery rate.

of the 32 replicated 500-individual populations with labeled, virgin hermaphrodites of any strain, whereas with labeled males the marker was lost in only 8 out of 32 and with labeled mated hermaphrodites in 6 out of 32 cases. For the population size of 100, the marker was completely lost in 9 out of 12 replicated populations with virgin hermaphrodites, 6 out of 12 with males and 1 out of 12 with mated hermaphrodites. Nevertheless, our results illustrate that rare males in a hermaphroditic population cause an increase in frequency of their alleles in the next generations. This is obviously also the case for genes involved in male formation and development, which are necessarily functional if they occur in a male. If the occasional boost of frequency of functional alleles of these genes caused by the sporadic males is large enough to offset the loss of functional alleles by mutational degradation and drift (which is expected to happen in hermaphrodites), this might be sufficient to maintain the genetic machinery for the production of males, even if there is no fitness advantage of out-crossing for hermaphrodites as has been proposed by Chasnov (2002). However, it might also be advantageous for an individual hermaphrodite to produce males, by allowing X-chromosome non-disjunctions or by mating, as long as the hermaphrodite density is high and the frequency of males is low, because this should indirectly lead to an increase of the frequency of the hermaphrodite's alleles. This advantage is expected to be strongly enhanced if novel environmental conditions (i.e. new selective constraints) can be expected to favor new phenotypes, because such new phenotypes are more rapidly produced through out-crossing and recombination than a series of mutations (Maynard-Smith 1978; Bell 1982). This notion recently was supported experimentally (Morran et al., 2009b).

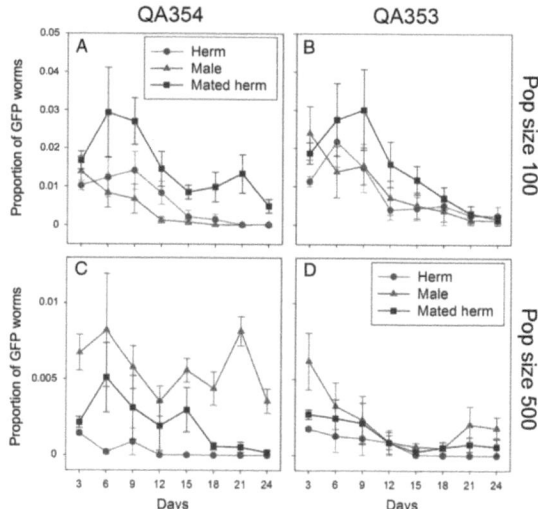

FIG. 2. Invasion Experiment 2: Proportion of GFP-positive worms (Y-axis) over time (X-axis) after the addition of a single GFP marked virgin hermaphrodite (circles), male (triangles) or mated hermaphrodite (squares) to a population of 500 (A, B) or 100 (C, D) N2 worms on day 0. The added worms were of strain QA354 (essentially the genetic background of CB4856; A, C) or QA353 (essentially the genetic background of N2; B, D). The error bars designate standard errors. Each point is the average of 6 independent measurements. For the statistical analysis see Table 2.

TABLE 2. Statistical analysis of the effect of treatment on the proportion of GFP-positive offspring in Invasion Experiment 2.

Cross	Population size	Comparison	$\chi^2_{df=1}$	P
QA354	100	Overall model	60.3	< 0.0001[a]
		Herm. vs. Male	4.5	0.0347[a]
		Herm vs. Mated herm.	32.5	< 0.0001[a]
		Male vs. Mated herm.	58.0	< 0.0001[a]
QA354	500	Overall model	182.7	< 0.0001[a]
		Herm. vs. Male	133.6	< 0.0001[a]
		Herm vs. Mated herm.	96.8	< 0.0001[a]
		Male vs. Mated herm.	53.4	< 0.0001[a]
QA353	100	Overall model	12.0	0.0025[a]
		Herm. vs. Male	1.8	0.1762
		Herm vs. Mated herm.	18.5	< 0.0001[a]
		Male vs. Mated herm.	4.8	0.0285[a]
QA353	500	Overall model	51.5	< 0.0001[a]
		Herm. vs. Male	54.4	< 0.0001[a]
		Herm vs. Mated herm.	20.1	< 0.0001[a]
		Male vs. Mated herm.	11.7	0.0006[a]

All models were significantly better than a minimal model ($\chi^2_{df=8-10} \geq 48.0$, $P < 0.0001$). The impact of the treatment factor was evaluated with a likelihood ratio effect test.

[a] Significant probabilities (P) according to the false-discovery rate.

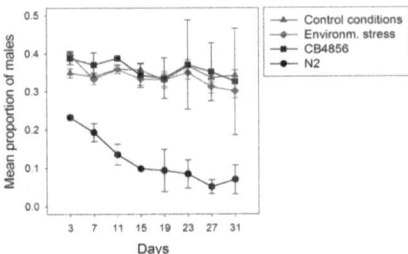

FIG. 3. Persistence of males over time in populations after the Experimental Evolution Experiment. The mean proportion of males (Y-axis) over time (X-axis) is given. Every four days the populations were reduced to 150 individuals and transferred to new plates. The error bars indicate standard errors. All experiments were started with populations containing approximately 50% males. The first actual measurement was done after the first generation at day 3. For the experimental (diamonds) and the control (triangles) treatments each point is the average of two independent measurements for each of the 10 replicates of the selection experiment (total of 20 data points per treatment and time point). Two independent, male maintenance assays for each of N2 (circles) and CB4856 (squares) were done in parallel as experimental controls. The average of these two measurements is shown. For the statistical analysis see Table 3.

Our prior work has shown that under standard laboratory conditions, different strains of *C. elegans* lose or maintain males at very different rates and levels (Wegewitz et al., 2008). This indicates that male maintenance is, at least in part, genetically determined and is therefore a selectable trait. In order to address the question if changing environmental stress conditions represent a selective pressure in favor of more outcrossing, we subjected worm populations to alternating conditions of high salt, low temperature, high temperature, and pathogenic bacteria or permanent standard laboratory conditions as control (Experimental Evolution Experiment). The conditions were changed every four days, which is slightly longer than one generation-time under standard laboratory conditions. Genetically heterogeneous starting populations were generated by interbreeding the strains N2 (low male maintenance) and CB4856 (high male maintenance). After five cycles of selection (corresponding to about 20 generations), we analyzed the male maintenance of the resulting populations under standard laboratory conditions (Fig. 3, Table 3). There was no difference between the selection and the control treatments. However, both control and selection populations had adopted a high male maintenance, which was significantly different from N2 but indistinguishable from CB4856.

Since during the Experimental Evolution Experiment small numbers of males of both parental strains were added to prevent stochastic loss of genotypes, we suspected that this might have led to the dominance of the CB4856-like phenotype. To address this, we repeated the control experiment in a simplified form, without the periodic addition of males during the experiment (Chunking Experiment). Again, all heterogeneous populations assumed a high male maintenance significantly different from N2 and undistinguishable from CB4856 (Fig. 4, Table 4). This result indicates that under standard laboratory conditions, subpopulations that behave like CB4856 with respect to male maintenance are selected for from N2 x CB4856 hybrid populations, at least if males are present in the cultures. The higher maintenance of males itself does not need to be the selected trait but it may be the consequence of selection for the CB4856 variant at closely linked loci. The high male maintenance might also be independent of the environment and result from intrinsic factors. Male maintenance may be influenced by a fairly large number of loci, and only if all or most of them are

TABLE 3. Statistical analysis of the effect of treatment/strains on the proportion of males in the Experimental Evolution Experiment.

Comparison	$\chi^2_{df=1}$	P
Overall model	108.9	< 0.0001[a]
Environm. stress vs. Control	0.4	0.5146
Environm. stress vs. CB4856	0.5	0.4693
Environm. stress vs. N2	92.2	< 0.0001[a]
Control vs. CB4856	0.2	0.6705
Control vs. N2	102.5	< 0.0001[a]
CB4856 vs. N2	49.7	< 0.0001[a]

With one exception, all models were significantly better than a minimal model ($\chi^2_{df=19-25} \geq 44.6$, $P < 0.0001$). The exception referred to the comparison between control conditions and CB4856, where the model only explained an insignificant part of the variance ($\chi^2_{df=12} = 16.8$, $P = 0.1584$), but still provided a good fit ($P > 0.999$); this comparison was still included to provide a complete overview of pairwise compared treatments/strains. The impact of the treatment factor was evaluated with a likelihood ratio effect test.

[a] Significant probabilities (P) according to the false-discovery rate.

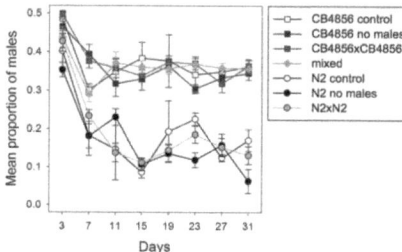

FIG. 4. Persistence of males over time in populations after the Chunking Experiment. The mean proportion of males (Y-axis) over time (X-axis) is given. Every four days the populations were reduced to 150 individuals and transferred to new plates. The error bars indicate standard errors. All experiments were started with populations containing approximately 50% males. The first actual measurement was done after the first generation at day 3. For the experimental and the control treatments each point is the average of two independent measurements for each of the 3 replicates of the Chunking Experiment. Two independent, male maintenance assays for each of N2 and CB4856 were done in parallel as experimental controls. The average of these two measurements is shown. CB4856 control (white squares); treatment CB4856 no males, (black squares); treatment CB4856 with males (grey squares); treatment mixed (grey diamonds); N2 control (white circles); treatment N2 no males (black circles); treatment N2 with males (grey circles).

N2-derived, low male maintenance occurs. If this is the case, the likelihood of recreating this situation from a mixed population by chance is very small. The low male maintenance in the standard laboratory strain N2 might then be the result of decade-long selection by geneticists, who prefer their strains to self-reproduce, unless mated deliberately. Alternatively, the relevant locus might reside in a region of genetic incompatibility between the two strains, such that preferentially individuals survive, which are homozygous for CB4856

TABLE 4. Significant results for the statistical analysis of treatment effects on the proportion of males during the Chunking Experiment.

Comparison	χ^2	P
Overall model	216	< 0.0001
Mixed vs. N2 control	30.2	< 0.0001
Mixed vs. N2 no males	53.7	< 0.0001
Mixed vs. N2 with males	66.6	< 0.0001
CB4856 control vs. N2 control	22.8	< 0.0001
CB4856 control vs. N2 no males	48.3	< 0.0001
CB4856 control vs. N2 with males	34.3	< 0.0001
CB4856 no males vs. N2 control	38.3	< 0.0001
CB4856 no males vs. N2 no males	83.3	< 0.0001
CB4856 no males vs. N2 with males	70.3	< 0.0001
CB4856 with males vs. N2 control	43.0	< 0.0001
CB4856 with males vs. N2 no males	91.3	< 0.0001
CB4856 with males vs. N2 with males	76.3	< 0.0001

The models of all shown cases were significantly better than a minimal model ($\chi^2_{df=5,22} \geq 25.1$, $P \ll 0.0001$). The impact of the treatment factor was evaluated with a likelihood ratio effect test. The degrees of freedom for the overall model is df = 6 and for all other comparisons df = 1. The probabilities (P) of all shown comparisons are significant according to the false-discovery rate.

derived genetic material in the particular region. We consider this explanation rather unlikely. These two strains are used extensively in the *C. elegans* field for genetic mapping and QTL analysis. Nevertheless, only one single region of partial genetic incompatibility between these two strains was found (Seidel et al., 2008). It affects an interval on chromosome I and clearly favors the N2-derived genetic material in this region.

It still remains to be addressed in the future whether higher outcrossing rates generally allow a population to adapt to new selective constraints more rapidly, as the recent findings suggest for pathogens (Morran et al., 2009b), and if this effect is indeed the decisive driving force behind the continuous existence of males in *C. elegans*.

LITERATURE CITED

Agrawal, A. F., and Lively, C. M. 2001. Parasites and the evolution of self-fertilization. Evolution: International Journal of Organic Evolution 55:869–879.

Barriere, A., and Felix, M. A. 2005. High local genetic diversity and low outcrossing rate in *Caenorhabditis elegans* natural populations. Current Biology 15:1176–1184.

Barriere, A., and Felix, M. A. 2007. Temporal dynamics and linkage disequilibrium in natural *Caenorhabditis elegans* populations. Genetics 176:999–1011.

Bell, G. 1982. Masterpiece of Nature: The evolution of sexuality. Berkeley: University of California Press.

Charlesworth, B., and Charlesworth, D. 1998. Some evolutionary consequences of deleterious mutations. Genetica 102–103:3–19.

Charnov, E. L., Smith, J. M., and Bull, J. J. 1976. Why be an hermaphrodite?. Nature 263:125–126.

Chasnov, J. R., and Chow, K. L. 2002. Why are there males in the hermaphroditic species *Caenorhabditis elegans*? Genetics 160:983–994.

Cutter, A. D. 2005. Mutation and the experimental evolution of outcrossing in *Caenorhabditis elegans*. Journal of Evolutionary Biology 18:27–34.

Cutter, A. D. 2006. Nucleotide polymorphism and linkage disequilibrium in wild populations of the partial selfer *Caenorhabditis elegans*. Genetics 172:171–184.

Cutter, A. D., Aviles, L., and Ward, S. 2003. The proximate determinants of sex ratio in *C. elegans* populations. Genetics Research 81:91–102.

Fischer, R. A. 1930. The genetical theory of natural selection. Oxford: Clarendon Press.

Gu, T., Orita, S., and Han, M. 1998. *Caenorhabditis elegans* SUR-5, a novel but conserved protein, negatively regulates LET-60 Ras activity during vulval induction. Molecular and Cellular Biology 18:4556–4564.

Haber, M., Schungel, M., Putz, A., Muller, S., Hasert, B., and Schulenburg, H. 2005. Evolutionary history of *Caenorhabditis elegans* inferred from microsatellites: evidence for spatial and temporal genetic differentiation and the occurrence of outbreeding. Molecular biology and evolution 22:160–173.

Hamilton, W. D., Axelrod, R., and Tanese, R. 1990. Sexual reproduction as an adaptation to resist parasites (a review). Proceedings of the National Academy of Sciences of the United States of America 87:3566–3573.

Hope, I. A. 1999. Background on *Caenorhabditis elegans*. Pp. 1–15 in I. A. Hope, ed. *C. elegans* a practical approach. Oxford: Oxford University Press.

Jin, Y. 1999. Transformation. Pp. 69–96 in I. A. Hope, ed. *C. elegans* a practical approach. Oxford: Oxford University Press.

Kiontke, K., Gavin, N. P., Raynes, Y., Roehrig, C., Piano, F., and Fitch, D. H. 2004. *Caenorhabditis* phylogeny predicts convergence of hermaphroditism and extensive intron loss. Proceedings of the National Academy of Sciences of the United States of America 101:9003–9008.

Kondrashov, A. S. 1988. Deleterious mutations and the evolution of sexual reproduction. Nature 336:435–440.

LaMunyon, C. W., and Ward, S. 1998. Larger sperm outcompete smaller sperm in the nematode *Caenorhabditis elegans*. Proceedings of the Royal Society - Biological Sciences 265:1997–2002.

LaMunyon, C. W., and Ward, S. 1999. Evolution of sperm size in nematodes: sperm competition favours larger sperm. Proceedings of the Royal Society - Biological Sciences 266:263–267.

LaMunyon, C. W., and Ward, S. 2002. Evolution of larger sperm in response to experimentally increased sperm competition in *Caenorhabditis elegans*. Proceedings of the Royal Society - Biological Sciences 269:1125–1128.

Loewe, L., and Cutter, A. D. 2008. On the potential for extinction by Muller's ratchet in *Caenorhabditis elegans*. BMC Evolutionary Biology 8:125.

Lopes, P. C., Sucena, E., Santos, M. E., and Magalhaes, S. 2008. Rapid experimental evolution of pesticide resistance in *C. elegans* entails no costs and affects the mating system. PLoS ONE 3:e3741.

Manoel, D., Carvalho, S., Phillips, P. C., and Teotonio, H. 2007. Selection against males in *Caenorhabditis elegans* under two mutational treatments. Proceedings of the Royal Society - Biological Sciences 274:417–424.

Mayer, W. E., Herrmann, M., and Sommer, R. J. 2007. Phylogeny of the nematode genus *Pristionchus* and implications for biodiversity, biogeography and the evolution of hermaphroditism. BMC Evolutionary Biology 7:104.

Maynard-Smith, J. 1978. The evolution of sex. Cambridge: Cambridge University Press.

Morran, L. T., Cappy, B. J., Anderson, J. L., and Phillips, P. C. 2009a. Sexual Partners for the Stressed: Facultative Outcrossing in the Self-Fertilizing Nematode *C. elegans*. Evolution; international journal of organic evolution 63:1473–1482.

Morran, L. T., Parmenter, M. D., and Phillips, P. C. 2009b. Mutation load and rapid adaptation favour outcrossing over self-fertilization. Nature 462:350–352.

Muller, H. J. 1932. Some genetic aspects of sex. American Naturalist 66:118–138.

Muller, H. J. 1964. The relation of recombination to mutational advantage. Mutation Research 1:2–9.

Pannell, J. R. 2002. The evolution and maintenance of Androdioecy. Annual Review of Ecology and Systematics 33:397–425.

Seidel, H. S., Rockman, M. V., and Kruglyak, L. 2008. Widespread genetic incompatibility in *C. elegans* maintained by balancing selection. Science 319:589–594.

Sivasundar, A., and Hey, J. 2003. Population genetics of *Caenorhabditis elegans*: The paradox of low polymorphism in a widespread species. Genetics 163:147–157.

Sivasundar, A., and Hey, J. 2005. Sampling from natural populations with RNAi reveals high outcrossing and population structure in *Caenorhabditis elegans*. Current Biology 15:1598–1602.

Stewart, A. D., and Phillips, P. C. 2002. Selection and maintenance of androdioecy in *Caenorhabditis elegans*. Genetics 160:975–982.

Stiernagle, T. 1999. Maintenance of *C. elegans*. Pp. 51–67 in I. A. Hope, ed. *C. elegans* a practical approach. Oxford: Oxford University Press.

Teotonio, H., Manoel, D., and Phillips, P. C. 2006. Genetic variation for outcrossing among *Caenorhabditis elegans* isolates. Evolution: International Journal of Organic Evolution 60:1300–1305.

Wegewitz, V., Schulenburg, H., and Streit, A. 2008. Experimental insight into the proximate causes of male persistence variation among two strains of the androdioecious *Caenorhabditis elegans* (Nematoda). BMC Ecology 8:12.

Wood, W. B. 1988. Introduction to *C. elegans* Biology. Pp. 1–16 in W. B. Wood, ed. The Nematode *Caenorhabditis elegans*. New York: Cold Spring Harbor Laboratory Press.

i want morebooks!

Buy your books fast and straightforward on ine - at one of world's fastest growing online book stores! Environmentally sound due to Print-on-Demand technologies.

Buy your books online at
www.get-morebooks.com

Kaufen Sie Ihre Bücher schnell und unkompliziert online – auf einer der am schnellsten wachsenden Buchhandelsplattformen weltweit! Dank Print-On-Demand umwelt- und ressourcenschonend produziert.

Bücher schneller online kaufen
www.morebooks.de

 VDM Verlagsservicegesellschaft mbH
Heinrich-Böcking-Str. 6-8   Telefon: +49 681 3720 174   info@vdm-vsg.de
D - 66121 Saarbrücken   Telefax: +49 681 3720 1749   www.vdm-vsg.de

Printed by Books on Demand GmbH, Norderstedt / Germany